EMOTIONAL INTELLIGENCE FOR LEADERSHIP EFFECTIVENESS

*Management Opportunities and
Challenges during Times of Crisis*

EMOTIONAL INTELLIGENCE FOR LEADERSHIP EFFECTIVENESS

Management Opportunities and Challenges during Times of Crisis

Edited by
Mubashir Majid Baba, PhD
Chitra Krishnan, PhD
Fatma Nasser Al-Harthy, PhD

First edition published 2023

Apple Academic Press Inc.
1265 Goldenrod Circle, NE,
Palm Bay, FL 32905 USA

760 Laurentian Drive, Unit 19,
Burlington, ON L7N 0A4, CANADA

CRC Press
6000 Broken Sound Parkway NW,
Suite 300, Boca Raton, FL 33487-2742 USA

4 Park Square, Milton Park,
Abingdon, Oxon, OX14 4RN UK

© 2023 by Apple Academic Press, Inc.

Apple Academic Press exclusively co-publishes with CRC Press, an imprint of Taylor & Francis Group, LLC

Reasonable efforts have been made to publish reliable data and information, but the authors, editors, and publisher cannot assume responsibility for the validity of all materials or the consequences of their use. The authors, editors, and publishers have attempted to trace the copyright holders of all material reproduced in this publication and apologize to copyright holders if permission to publish in this form has not been obtained. If any copyright material has not been acknowledged, please write and let us know so we may rectify in any future reprint.

Except as permitted under U.S. Copyright Law, no part of this book may be reprinted, reproduced, transmitted, or utilized in any form by any electronic, mechanical, or other means, now known or hereafter invented, including photocopying, microfilming, and recording, or in any information storage or retrieval system, without written permission from the publishers.

For permission to photocopy or use material electronically from this work, access www.copyright.com or contact the Copyright Clearance Center, Inc. (CCC), 222 Rosewood Drive, Danvers, MA 01923, 978-750-8400. For works that are not available on CCC please contact mpkbookspermissions@tandf.co.uk

Trademark notice: Product or corporate names may be trademarks or registered trademarks and are used only for identification and explanation without intent to infringe.

Library and Archives Canada Cataloguing in Publication

Title: Emotional intelligence for leadership effectiveness : management opportunities and challenges during times of crisis / edited by Mubashir Majid Baba, PhD, Chitra Krishnan, PhD, Fatma Nasser Al-Harthy, PhD.
Names: Baba, Mubashir Majid, editor. | Krishnan, Chitra, 1983- editor. | Al Harthi, Fatma, 1972- editor.
Description: First edition. | Includes bibliographical references and index.
Identifiers: Canadiana (print) 20220451877 | Canadiana (ebook) 20220451923 | ISBN 9781774911327 (hardcover) | ISBN 9781774911334 (softcover) | ISBN 9781003303862 (ebook)
Subjects: LCSH: Leadership. | LCSH: Emotional intelligence. | LCSH: Crisis management. | LCSH: Organizational behavior.
Classification: LCC HD57.7 .E46 2023 | DDC 658.4/092—dc23

Library of Congress Cataloging-in-Publication Data

....

CIP data on file with US Library of Congress

....

ISBN: 978-1-77491-132-7 (hbk)
ISBN: 978-1-77491-133-4 (pbk)
ISBN: 978-1-00330-386-2 (ebk)

About the Editors

Mubashir Majid Baba, PhD

Department of Management Studies, University of Kashmir, Jammu and Kashmir, India

Mubashir Majid Baba, PhD, is presently a faculty member in the Department of Management Studies, North Campus, at the University of Kashmir, Jammu and Kashmir, India. He previously worked as an Assistant Professor in the Department of Humanities, Social Sciences and Management, NIT Srinagar, and as an Assistant Professor in the Department of Management Studies, Cluster University of Srinagar (Jammu and Kashmir). Dr. Baba has two books to his credit and has published papers in journals of national and international (Scopus/SSCI indexed) journals and has presented papers at national and international conferences. He has also attended many workshops, seminars, and conferences. He has remained associated with the Indian Institute of Management, Lucknow (IIML) as a Project Fellow. He was selected for a prestigious Stipendium Hungaricum Scholarship in 2015 by the Tempus Public Foundation and University Grants Commission (UGC) for the PhD program at the University of Kosovar (a doctoral school for management and organization science), Hungary. Dr. Baba is a columnist for various newspapers, focusing on education. His areas of interest include organizational behavior, human resource management, leadership, and emotional intelligence. Dr. Baba holds a PhD in Management from the University of Kashmir and a master's degree in Business Administration (MBA) from the Central University of Kashmir, India. In addition, he holds a postgraduate diploma in International Business Operations (PGDIBO) and a postgraduate diploma in Information Technology (PGDIT).

Chitra Krishnan, PhD

Associate Professor of Human Resource Management & Behavioural Science, Symbiosis Centre for Management Studies, Noida (Constituent of Symbiosis International) (Deemed University), India

Chitra Krishnan, PhD, is an Associate Professor of Human Resource Management & Behavioural Science at the Symbiosis Centre for

Management Studies, Noida, India. She is a teaching professional with over 16 years of national and International experience. She possesses excellence in teaching and research. Before her academic career, she has worked in industry in various positions of responsibility. She has been actively involved in rigorous academic pursuits in the field of higher professional education to enhance skill sets that promote holistic development of learners. She has a number of publications in acclaimed journals at national and international level and has also participated in many national and international conferences. She is passionate toward writing and has five books to her credit with international publishers. She has been empanelled as a member of review committees for conferences and journals of repute. Her area of interest includes human resource management, organisation behaviour, talent management, diversity management, employee satisfaction, knowledge management, artificial intelligence, and emotional intelligence.

Fatma Nasser Al-Harthy, PhD

Assistant Dean for Students' Affairs,
University of Technology and Applied Science (UTAS), IBRA, Oman

Fatma Nasser Al-Harthy, PhD, is working at the University of Technology and Applied Science (UTAS), IBRA, Oman, as an Assistant Dean for Students' Affairs. Her research interests are mainly in field of leadership, management, women's studies, accounting and finance, and entrepreneurship and innovation. She has supervised research at The Research Council, Oman, under the Faculty Mentored Undergraduate Research Award Program. Dr. Fatma is a member of several research programs in Oman, such as the Directorate General of Health in Sharqia Region Committee and the HR Research Agenda of technical education, Oman. Dr. Fatma has published various papers in Scopus journals as well as in proceedings of international conferences. She has held several workshops regarding how to write a good literature review for UTAS business students and delivered workshops on leadership to other higher educational institutions in Oman. She was awarded her PhD from the University Sains Malaysia (USM) and her MBA from Victoria University, Australia.

Contents

Contributors .... *ix*

Abbreviations .... *xiii*

Foreword .... *xv*

Introduction .... *xvii*

Preface .... *xix*

1. **Factors Affecting Psychological Needs of Medical Workforce During the COVID-19 Situation** 1

   Hardeep Singh, Ruchi Tyagi, Sumeet Gupta, and Anurag Sharma

2. **Impact of COVID-19 on Employees' Emotional Health and the Role of Leadership** 17

   Mansi Babbar and Ishita Khanna

3. **The Impact of Emotional Intelligence and Leadership in Pandemic Times** 37

   Mariana Rezende Alves de Oliveira, Mubashir Majid Baba, and José Aparecido Da Silva

4. **Emotional Intelligence and Its Management in the Conflicting Factors in the Pandemic by Using a Mediative Fuzzy Logic System** ... 53

   Nitesh Dhiman, Meetika Sharma, and M. K. Sharma

5. **Unlocking the Mask: A Look at the Role of Leadership and Innovation Amid the COVID-19 Pandemic Crisis** 81

   Sabzar Ahmad Peerzadah, Sabiya Mufti, and Shayista Majeed

6. **Emotional Intelligence as a Tool to Manage Conflict, Emotions, and Behavior of Human Beings During the Pandemic COVID-19** 99

   Shruti Traymbak, Meghna Sharma, Shubham Aggarwal, and Krity Gulati

7. **Thriving in the "New Normal": An Era of COVID-19** 107

   Mohd. Zia Ul Haq Rafaqi and Zainab Musheer

viii   *Contents*

8. Depression, Anxiety, and Stress of Coronavirus-Infected People of Kashmir in Relation to Psychological Hardiness 121

Shabir Ahmad Malik

9. Self-Management During the COVID-19 Crises 131

Rehana Amin and Mohammad Maqbool Dar

10. Theoretical Framework on the Need of Emotional Balance and Work-Life Balance for Employees in the Epidemic Situation: A Reference to COVID-19 141

Neelni Giri Goswami

11. Managing Emotions in the Pharmaceutical Sector: How to Minimize Negative Emotions 159

Soumendra Darbar, Srimoyee Saha, and Sangita Agarwal

12. Emotions, Perceptions, and the Organizational Dynamism Required in Tourism During a Crisis 173

Zepphora Lyngdoh

13. Are Organizations Ready to Manage Stress During and After the COVID-19 Pandemic? 191

Rajni Singh

14. Apprehension of Organization Culture Through Emotional Stability 201

Garima Saini and Shabnam

15. Emerging Ethical Leadership in Crisis Management 221

Pavitra Dhamija

Index *237*

Contributors

Sangita Agarwal
Department of Applied Science, RCC Institute of Information Technology, Canal South Road, Beliaghata, Kolkata–700015, West Bengal, India

Shubham Aggarwal
Department of Management, Lloyd Business School, Greater Noida, Uttar Pradesh, India

Rehana Amin
Faculty of Department of Psychiatry, Government Medical College Srinagar (Institute of Mental Health and Neurosciences–Kashmir), Jammu and Kashmir, India, E-mail: drrehanaamin2@gmail.com

Mubashir Majid Baba
Department of Management Studies, North Campus, University of Kashmir, Srinagar, Jammu and Kashmir, India; Former Project Fellow, IIM Lucknow, Uttar Pradesh, India

Mansi Babbar
Department of Commerce, Shaheed Bhagat Singh College, University of Delhi, Delhi–110017, India, E-mail: mansibabbar21@gmail.com

Mohammad Maqbool Dar
Faculty of Department of Psychiatry, Government Medical College Srinagar (Institute of Mental Health and Neurosciences–Kashmir), Jammu and Kashmir, India

Soumendra Darbar
Research and Development Division, Dey's Medical Stores (Mfg.) Ltd., 62, Bondel Road, Kolkata–700019, West Bengal, India, E-mail: dr.soumendradarbar@deysmedical.com

Pavitra Dhamija
LM Thapar School of Management, Thapar Institute of Engineering and Technology, Patiala (Punjab), India; CIDB Centre of Excellence, Department of Construction Management and Quantity Surveying, Faculty of Engineering and the Built Environment, University of Johannesburg, Johannesburg, South Africa, E-mail: pavitradhamija@gmail.com

Nitesh Dhiman
Department of Mathematics, Chaudhary Charan Singh University, Meerut–250004, Uttar Pradesh, India

Neelni Giri Goswami
Assistant Professor, University School of Business-Commerce, Chandigarh University, Chandigarh, India, E-mail: neelnigoswami@gmail.com

Krity Gulati
Department of Management, LIMT, Greater Noida, Uttar Pradesh, India

Sumeet Gupta
Department of General Management, School of Business, University of Petroleum and Energy Studies, Dehradun, Uttarakhand, India

Ishita Khanna
University of Delhi, Delhi–110017, India

Zepphora Lyngdoh
Department of Tourism and Hotel Management, NEHU, Shillong–793022, Meghalaya, India,
E-mail: zeyadkhar@gmail.com

Shayista Majeed
School of Business Studies University of Kashmir, Srinagar, Jammu and Kashmir–190006, India

Shabir Ahmad Malik
Research Scholar, Department Teacher Training and Non-Formal Education, IASE,
Faculty of Education, Jamia Millia Islamia University, New Delhi, India,
E-mail: malikshabir622@gmail.com

Sabiya Mufti
School of Business Studies University of Kashmir, Srinagar, Jammu and Kashmir–190006, India

Zainab Musheer
Department of Education, Aligarh Muslim University, Aligarh, Jammu and Kashmir, India

Mariana Rezende Alves de Oliveira
Master's Student in Psychobiology–FFCLRP, USP–University of São Paulo at Ribeirão Preto,
Virtual Laboratory of Affective, Cognitive, and Behavioral Neuropsychometry–LAVINACC,
E-mail: mari-rao@hotmail.com

Sabzar Ahmad Peerzadah
School of Business Studies University of Kashmir, Srinagar, Jammu and Kashmir–190006, India

Mohd. Zia Ul Haq Rafaqi
Department of Education University of Kashmir, South Campus, Anantnag, Jammu and Kashmir,
India, E-mail: mzrafiqi@gmail.com

Srimoyee Saha
Department of Physics, Jadavpur University, 188, Raja S.C. Mallick Road, Kolkata–700032,
West Bengal, India

Garima Saini
Research Scholar, Department of Humanities and Social Sciences, National Institute of Technology,
Kurukshetra, Haryana, India, E-mail: Garimasaini3@gmail.com

Shabnam
Assistant Professor, Department of Humanities and Social Sciences, National Institute of Technology,
Kurukshetra, Haryana, India

Anurag Sharma
Department of General Management, School of Business, University of Petroleum and Energy
Studies, Dehradun, Uttarakhand, India

M. K. Sharma
Department of Mathematics, Chaudhary Charan Singh University, Meerut–250004, Uttar Pradesh,
India, E-mail: drmukeshsharma@gmail.com

Meetika Sharma
Department of English, J.L.M. College, OFC, Muradnagar, Ghaziabad–201206, Uttar Pradesh, India

Contributors

Meghna Sharma
AIBS, Amity University, Noida, Uttar Pradesh, India

José Aparecido Da Silva
Full Professor of 4P: Psychometrics, Psychophysics, Perception,
and Pain Department of Psychology, University of São Paulo at Ribeirão Preto, Brazil

Hardeep Singh
Department of General Management, School of Business, University of Petroleum and Energy Studies,
Dehradun, Uttarakhand, India

Rajni Singh
Department of Management, Hierank Business School, Noida, Abdul Kalam Technical University
(AKTU), Uttar Pradesh, India, E-mail: rajni.singh2009@rediffmail.com

Shruti Traymbak
Department of Management, Lloyd Business School, Greater Noida, Uttar Pradesh, India,
E-mail: shruti@lloydcollege.in

Ruchi Tyagi
Department of General Management, School of Business, University of Petroleum and Energy
Studies, Dehradun, Uttarakhand, India, E-mail: rtyagi@ddn.upes.ac.in

Abbreviations

Apps	applications
AVP	assistant vice president
BMW	Bayerische Motoren Werke AG
BRS	brief resilience scale
CCC	COVID-19 care centers
CEO	chief executive officer
CG	caregiver
CI	confidence interval
CNE	common negative emotions
COR	conservation of resources
CoV	coronavirus
COVID-19	coronavirus disease 2019
DASS	depression anxiety stress scale
Div	division
E	education loss
EI	emotional intelligence
EQ	emotional quotient
ER	emotion regulation
ESM	experience sampling method
ETC	etcetera
ETX	embedded technology extended
F	financial crisis
FHBR	fortune and Harvard business review
FTSE	financial times stock exchange
GAS	general adaptation syndrome
HCL	Hindustan Computers Limited
HR	human resource
HRIS	human resources information system
HRM	human resource management
I	interpersonal effects
IANS	Indo-Asian News Service
IASE	Institute of Advanced Studies in Education
ICC	intra class correlation coefficient

IFTF	Institute for the Future
IQ	intelligence quotient
IRPFS	Indian Railway Protection Force Service
IT	information technology
K6	6-item Kessler psychological distress scale
LDCs	less developed countries
LQ	learnability quotient
LS	life satisfaction
MERS	Middle East respiratory syndrome
nCoV	novel coronavirus
NFE	non-formal education
OCD	obsessive-compulsive disorder
OEA	other's emotional appraisal
OHQ	Oxford happiness questionnaire
OMG	Omnicom Media Group
P	psychological behavior
PAN	presence across nation
PEC	positive emotional climate
PM	positive emotion
Pvt.	private
ROE	regulations of emotion
RQDA	R package for computer assisted qualitative data analysis
SARS	severe acute respiratory syndrome
SARS-CoV-2	severe acute respiratory syndrome coronavirus 2
SEA	self-emotion appraisal
Sec	section
SERS	severe acute respiratory syndrome
SOP	standard operating procedure
SWLS	satisfaction with life scale
TL	transformational leadership
TT	teacher training
TV	television
UOE	use of emotion
WHO	World Health Organization
WTO	World Tourism Organization
YOGA	your objectives, guidelines, and assessment

Foreword

The Fourth Industrial Revolution (4IR) is characterized by the fusion of the digital and physical worlds with the growing utilization of new technologies such as artificial intelligence, cloud computing, robotics, 3D printing, the Internet of Things, and advanced wireless technologies. These developments have ushered in a new era of economic disruption with uncertain socio-economic consequences. Therefore, the new reality of today's organization—globalization impacts, outsourcing, advanced technology, and virtual teams; and e-business—have caused uncertainty for people across the world. There are tremendous changes happening in the world that require fundamental demand shifts. It is necessary to move from an old paradigm to a new one. The COVID-19 pandemic has placed extraordinary demands on businesses and other leaders. Employees and other stakeholders are fearful because of COVID-19's humanitarian toll. Executives are finding it difficult to respond due to the outbreak's massive scale and sheer unpredictability. In fact, the outbreak is characterized by a crisis of "landscape-scale": an unexpected event or sequence of events of massive scale and extreme speed, causing high levels of uncertainty, disorientation, a sense of loss of control, and strong emotional disorder. Leaders can start to respond once they recognize a crisis. Yet in a routine emergency, they cannot react as they would, by following plans drafted in advance. Effective responses are improvised in large measure during a crisis that is governed by unknowingness and uncertainty. They could cover a variety of measures: not only temporary steps, but also changes to ongoing practices in businesses, which can be useful to continue even after the crisis has passed. During a crisis, what leaders require is not a predefined response plan, but rather behaviors and mindsets that prevent them from overreacting to yesterday's events and allow them to look ahead.

Under this backdrop, this is a commendable attempt by the editors and authors to write a this book, *Emotional Intelligence for Leadership Effectiveness: Management Opportunities and Challenges during Times of Crisis.* This edited book attempted to examines the problems faced due to crisis through the lens of emotional intelligence and leadership roles. I am very confident that this edited book will add value to the existing literature of this area and will be a must-read book to academics, researchers, and practitioners as well.

—Dr. Md. Aminul Islam
Faculty of Applied and Human Sciences,
University Malaysia Perlis, Malaysia

Introduction

The outbreak of COVID-19 has resulted in tension, anxiety, and fear, which has led to psychological disorders. The pandemic is considered one of the most turning points in history as it hobbles economic and social standards by generating a new human era. It is impossible to know what the new world will look like. The global COVID-19 pandemic has brought about a series of changes to the way we work. From suddenly managing teams working remotely to employees experiencing mental health or financial hardship – the crisis has led to many new leadership challenges.

Leadership may be difficult to describe, but in times of crisis, it is easy to identify. As the pandemic has spread fear, disease, and death, national leaders around the world have been severely challenged. Some leaders have failed miserably, but others have risen to the challenge, exhibiting resolve, courage, empathy, respect for science and essential decency, and thereby reducing the burden of the disease on their people. It is the responsibility of leaders to show empathy, optimism, and flexibility that will lead the people out of this crisis, which will ultimately be depicted in individuals as well.

Amid the chaos of recent months, the world has latched onto leaders who seem capable of righting the ship. Some we may have expected, such as heads of state like New Zealand's popular Prime Minister Jacinda Ardern and Germany's longtime Chancellor Angela Merkel. Others were perhaps less obvious, including NBA Commissioner Adam Silver, 3M CEO (Chief Executive Officer) Mike Roman, and New York Governor Andrew Cuomo. However, the phrases employed to define their leadership styles can be utilized to draw a relationship between all of these individuals. Silver's remark has been described as "honest" and Cuomo's as "reassuring." They each have a relational style to leadership that helps them to build a positive emotional climate (PEC) and gain buy-in for a shared vision of how to weather our current problems.

It is always important for leaders to have emotional intelligence (EI), but it becomes much more important during times of crisis. Emotions are high when things are unpleasant or tense. To be a leader, you must control your own emotions and help others do the same. The emotionally intelligent leader is more likely to successfully manage many relationships in a crisis. Inspiring others and managing conflict, for instance, is easier for those leaders who can connect on a deeper level via EQ. Leaders with high EQ know themselves. They can effectively self-regulate and self-motivate through a difficult, uncertain time. Coming from a solid self-foundation, these leaders are able to engage effectively with others. Many academics claim that leaders with emotional intelligence are better prepared to face the challenges of a crisis and are better able to build trust among employees, encourage innovation, and drive them more than those without it. The pandemic has presented the ultimate opportunity to assess leaders as they deal with unprecedented challenges that are testing their values and skills. There has never been a more extraordinary moment to strengthen and apply emotional intelligence competencies.

This book, *Emotional Intelligence for Leadership Effectiveness: Management Opportunities and Challenges during Times of Crisis*, combines through integrated chapterization on capabilities in the area of emotional intelligence (EI) that can help leaders to tackle any crisis with reduced stress levels, emotional restraint, and less unintended repercussions.

Preface

This book, *Emotional Intelligence for Leadership Effectiveness: Management Opportunities and Challenges during Times of Crisis,* elaborates on perspectives of emotional intelligence for the efficacy of leadership during the COVID-19 pandemic. In the book, you will certainly also discover the creative and intellectual impulses of the authors, which are brilliantly represented. The book contains a variety of emotional intelligence and leadership topics. Some of the key themes discussed are:

Chapter 1 provides an overview of the psychological influences of medical workers who provide services during COVID-19, which are anxiety, depression, and stress. The medical personnel require appropriate techniques to cope and sustain their mental well-being in these difficult times. In order to ensure emotional well-being in healthcare personnel, the government and politicians should institutionalize the care of medical workers.

Chapter 2 highlights the barriers and psychological interactions faced by employees during a pandemic of COVID-19. The chapter also highlights the elements that cause the pandemic, such as social distancing, financial difficulties, and the continuously spread negatives through social media. This highlights the responsibility of leaders in organizations, such that even in the face of a disaster to keep employees motivated.

Chapter 3 focuses on evaluating the factors that could have an effect on persons dealing with CoV-19-pandemic: emotional insight, happiness, psychological stress, resilience, and non-somatic pain. These considerations may have influenced the way that every national leader has decided to tackle the epidemic and its effect on the population's mental health.

Chapter 4 shows that how meditative fuzzy logic can be applicable in emotional intelligence to manage the conflicting factors during pandemic. Meditative fuzzy logic is an extension of intuitionistic fuzzy logic; it has the ability to control contradictory information. Meditative fuzzy logic can be helpful in the management of emotional quotient (EQ) factor, to develop the EQ skill level.

Chapter 5, in particular, discusses the pandemic role of leadership and creativity. Thus, it is the high leader and his imaginative answers to the problems that are the responsibility during a crisis. This chapter also highlights the steps that leaders must take to promote innovation.

Chapter 6 discusses four important dimensions of emotional intelligence: self-emotion appraisal (SEA), use of emotion (UOE), other's emotional appraisal and regulation of emotion (ROE), and their relationship with life satisfaction (LS). UOE refers to the use emotions and feelings in constructive way and regulation of emotions (ROE) helps an individual to cope up with psychological stress and both have a significant positive impact on life satisfaction. Interestingly, gender has no moderating effects between SEA, ROE, UOE, and OEA and life satisfaction.

Chapter 7 provides vital insight on the way in which college staff in India began the psychological recovery process, even if the risk of COVID-19 is still present in the environment. This chapter addresses the signs of endangered independence: impudence and decreased authenticity in the process of psychological reconstruction. During unlock—I and unlock—VI COVID-19, ESM design is employed for studying the psychological recovery process amongst university staff in India.

Chapter 8 concentrates on the psychological resilience of persons in Kashmir affected by coronavirus, the amount of depression, anxiety, and stress. The author concluded that coronavirus-contained persons need sufficient counseling to reduce their psychological impact in accordance with COVID-19. Increasing psychical strength in persons with coronavirus would also lessen depression, anxiety, and stress.

Chapter 9 highlights self-maintenance during the pandemic COVID-19. A person's social, emotional, and physical well-being can be retained by following self-care planning. The level of self-care determines an individual's competence and enables them to stand during crises. Practicing a healthy lifestyle, relationships, stress prevention, knowledge update are many self-care approaches, apart from consulting professionals, such as consultants, psychologists, and clinical psychiatrists, as required.

Chapter 10 focuses on the employee's perception of forced work-from-home initiative. This chapter helps businesses to understand the role of employees in policy and strategic development. Over time, the connection between work and life becomes lighter and eventually results in employees working emotionally, constantly, and happily.

Preface xxi

Chapter 11 focuses on emotions, behavior, social status, and overall workplace temperature during the pandemic. The section focuses on decision-making skills, team spirit, and leadership skills in lock-down and mental health consequences. This section presents an outline of how we handle emotion in the pharmaceutical industry and how negative emotions might be minimized.

Chapter 12 focuses on the peculiarities of the crisis situation and organizational intellectuality required during this dysfunction to diagnose issues and to manage tourism.

Chapter 13 highlights the need for organizations and individuals to establish certain methods to meet the difficulties of the pandemic. Some of the tactics examined in the study are: (i) communication transparency; (ii) reinforcement of the HR policies; (iii) COVID-19 impact mitigation training; (iv) social assistance and interactions with workers; and (v) preparation and communication of the right plan to physically return work. The current study will prepare organizations and people to deal with the pandemic's issues.

Chapter 14 proposes a roadmap that helps to build a continuous culture that helps to deliver long-term benefits for the organization by acknowledging efforts of all team members, encouragement of the voice of staff, priority for the values of businesses, positive, and strong team connections, continual education, and investments in staff development.

Chapter 15 emphasizes that during crisis management, ethical leadership has become one of the important factors. Implementing and using this principle sincerely can be very useful to organizations during pandemics. The different themes for ethical leadership emerge from the investigation, particularly when everything has actually become uncertain, are very relevant and important for crisis management. Crisis is a very huge example of how effective leadership can return to their original and normal condition.

We hope this book will greatly benefit academics, managers, strategists, politicians, and dynamic students who are the foundation of today's intelligentsia.

—Editors

CHAPTER 1

Factors Affecting Psychological Needs of Medical Workforce During the COVID-19 Situation

HARDEEP SINGH, RUCHI TYAGI, SUMEET GUPTA, and ANURAG SHARMA

Department of General Management, School of Business, University of Petroleum and Energy Studies, Dehradun, Uttarakhand, India, E-mail: rtyagi@ddn.upes.ac.in (R. Tyagi)

ABSTRACT

The sudden increase in the number of people infected with COVID-19 puts pressure on the medical workforce to treat the infected patients; the medical workforce, especially nurses and paramedical staff, is insufficient training to treat infected people. The lack of facility brings mental pressure to the medical workforce and brought significant psychological changes in their behavior. There was a scarcity of protective equipment available to the medical workforce that increased the risk of infection among healthcare providers to alleviate the issue. The medical workforce has experienced specific psychological changes due to COVID. This study considers a need as a psychological feature that arouses an organism to action towards a goal giving some purpose and direction to behavior. The objective of the present study is to identify the factors affecting the psychological needs of caregivers in disaster situations- COVID-19 and to identify the psychological needs that would lead to focused interventions ultimately reduces psychological stress-related disorders of health workers.

Emotional Intelligence for Leadership Effectiveness: Management Opportunities and Challenges During Times of Crisis. Mubashir Majid Baba, Chitra Krishnan, & Fatma Nasser Al-Harthy (Eds.) © 2023 Apple Academic Press, Inc. Co-published with CRC Press (Taylor & Francis)

1.1 SYNONYMS

Caregivers' (CG) psychological needs, medical workforce psychological needs.

1.2 DEFINITION

The psychological needs of nurses, medical workforce, and family caregivers who do caregiving to COVID-19 patients involve taking on practical supportive tasks. Caregiving involves caring for someone who has obvious emotional distress and connotations with patients.

1.3 INTRODUCTION

COVID-19 has affected the lives of billions of people around the globe, and healthcare providers have experienced changed psychological needs during COVID-19 [1]. Viral pandemics have affected all of us from time to time despite advancements in healthcare science. The 21st century has experienced an outbreak of severe acute respiratory syndrome (SARS-CoV), middle east respiratory syndrome (MERS-CoV), and Ebola [2]. COVID-19 is no exception in terms of negatively influencing the well-being of people at large. COVID-19 was earlier called 2019-nCoV, and the International Committee of Viral Classification renamed it COVID-19 on February 12 2020. Patients with SARS-CoV-2 (severe acute respiratory syndrome coronavirus 2) infection infect others due to COVID-19 infection. Older people, children, and infants are more likely to develop this disease. COVID-19 originated from an unexplained source from Wuhan, and it has spread all over the globe. COVID-19 has challenged the public health systems of nations, and it has become challenging for Governments to balance between social distancing and economic growth. States are struggling to find methods to gain economic momentum while maintaining social distancing. Considering COVID-19 on the entire economic system, the industries with lower variability and stable employment opportunities are reshaping drastically by COVID-19. Figure 1.1 highlights the increase in the COVID-19 cases in India and its neighborhood. WHO (World Health Organization) dashboard [23] information showing.

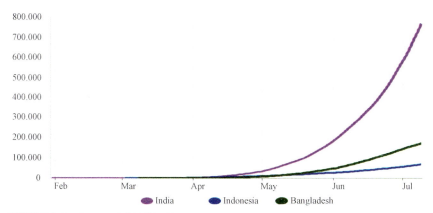

FIGURE 1.1 Count of COVID-19 cases.

The sudden increase in the number of people infected with COVID-19 puts pressure on the medical workforce to treat the infected patients; the medical workforce, especially nurses and paramedical staff, is insufficient training to treat infected people. The lack of facility brings mental pressure to the medical workforce and brought significant psychological changes in their behavior. There was a scarcity of protective equipment available to the medical workforce that increased the risk of infection among healthcare providers to alleviate the issue.

The medical workforce has experienced specific psychological changes due to COVID-19 [3–5]. This study considers a need as a psychological feature that arouses an organism to action towards a goal giving some purpose and direction to behavior. The psychological need provides the feeling of one's action with some form of pro-social impact that can potentially increase a person's experience of beneficence. Psychological needs have a direct effect on personal growth, psychological development, and self-actualization. The three basic psychological needs of caregivers identified in the literature are autonomy, competence, and relatedness. Human beings have three inborn basic needs: autonomy, competence, and relatedness.

The nutrients for psychological growth, integration, and mental well-being are psychological rather than physiological needs. Transactional analysis theory divides psychological needs into three parts: [1] hunger of structure; [2] hunger of stimuli; [3] hunger of strokes fulfilling at the personal and professional level or both.

Psychological needs are subjective experiences/feelings that we need to be happy and contented. The need for autonomy is a psychological need to feel a sense of freedom and have the opportunity. It is about experiencing positive emotions and a sense of physical health and mental well-being. Therefore, understanding the psychological needs of healthcare providers is crucial.

The objectives of the present study are:

- To identify the factors affecting the psychological needs of caregivers in disaster situations – COVID-19;
- Identifying the psychological needs that would lead to focused interventions ultimately reduces psychological stress-related disorders of health workers.

1.4   LITERATURE REVIEW

There is very little information available on caregiving service providers [6]. A higher level of anxiety, depression, and lower mental well-being during COVID-19 and the gender differences in anxiety, depression, and well-being are non-significant among genders is observed in the literature. People in the age group of 21 to 40 are more vulnerable to psychological problems [2]. Emphasizing more online campaigns to mitigate psychological problems and understanding psychological needs are necessary to solve psychological problems. The importance of protective well-being and the use of AI (artificial intelligence) tools to control COVID-19 to protect well-being is also crucial [7]. The development of applications to automate COVID-19 contact tracing is essential for psychological well-being [8]. The integration of technology claims to handle issues such as COVID-19 to maximize public benefits. Consideration of the user's privacy needs while launching such an application is vital so that the objective to endorse well-being can be achieved [9, 10]. The impact of COVID-19 on the psychology of healthcare workers is studied; findings suggest that due to COVID-19, healthcare workers were experiencing psychological distress [11]. Studies with a cross-sectional survey identified the participant's psychological status. The exposure of healthcare workers to the risk of being infected was a significant reason for psychological stress among healthcare workers. The requirement of early prevention of consequences of physical distancing and we have shown the same for the medical workforce [12]. There is a facility requirement at the national level for psychological first aid in times of pandemics. The medical workforce involved in interactions with the infected people reported a higher score on psychological disorders. They suggested designing effective strategies to improve the mental health of the medical workforce. The support of nurses by their teams and the public for the resources is essential during difficult times [13]. The requirement of protective equipment for nurses is necessary for their mental well-being. There is a requirement to take care of the well-being of the nurses during caregiving Nurses go through the emotional stress of pandemic, COVID-19. The studies pinpoint the need for support for the mental well-being of nurses after the COVID-19. A nationwide survey among Chinese on psychological distress during COVID-19 argued that COVID-19 triggered psychological issues such as disorder, anxiety, and depression [14]. The survey documents that during epidemics such as COVID-19, early recognition, and establishment of strategic plans to reduce the distress due to psychological needs for more attention to vulnerable groups, improving accessibility of medical resources, initiation of programs focusing on psychological aid is a must. The studies focused on the need for psychological needs assessment, and this study presents results on similar grounds. The literature review shows concern for mental health, highlighting that depression and anxiety are common psychological responses among healthcare and the common public. Consultation with the online mode is an effective tool for patients [15]. The emotional well-being of healthcare workers is also studied [15]. Three critical areas for the well-being of healthcare workers are meeting the workforce's basic needs, improving the quality of messages, and accessibility to psychological and mental support decisions during COVID-19 situation [16]. A longitudinal study of the general population during COVID-19 stated a significant reduction in stress during the initial four weeks during COVID-19 in China [17].

They suggested that the government ensure the availability of essential services, creating awareness about the disease, and providing financial support during COVID-19. Medical care workers' mental health and psychological problems during COVID-19 in China are also studied [18] and showed insomnia, anxiety, depression, somatization, and obsessive-compulsive symptoms to be among the psychological problems of medical health care. They argued for recovery problems for the mental well-being of medical health care workers because such workers experience psychological problems.

The present study has limitations. Firstly, the study is when COVID-19 cases were at peak during the first wave of the pandemic in India, and a higher prevalence of psychological symptoms can be among the medical workforce. Secondly, the study includes only the doctors and nurses and not pharmacists, psychologists, and administrative staff of hospitals. Because of this, the sample size is small but enough to give reliable findings.

1.5 RESEARCH METHODOLOGY

1.5.1 RESEARCH DESIGN

A qualitative research design is the approach of the present study. The information is collected from the medical workforce directly involved in the caregiving activities during COVID-19 in India. We conducted semi-structured interviews of the medical workforce to understand the psychological needs during the pandemic situation of COVID-19 in India. Semi-structured interviews were pre-planned for the objective of the present study, and the details of such interviews are present in the following sections.

1.5.2 SAMPLE SELECTION

We identified interviews using judgmental sampling methods and selected 15 medical workforces directly involved in treating COVID-19 infected patients (details given in Table 1.1). We selected people who had a minimum of two years of experience and selected only doctors and nurses who provide health care services and are in the age group of 24 years to 40 years. The selected participants are the most vulnerable group for psychological disorders.

1.5.3 RESEARCH INSTRUMENTS

As mentioned earlier, themes of depression, anxiety, and stress among the healthcare workforce were developed based on the literature review. The triangulation, including a pilot study on the semi-structured interview, validates the interview guide for collecting the required information from

Factors Affecting Psychological Needs of Medical Workforce 7

the healthcare workforce. The focus of the first theme is on understanding the status of the impact of COVID-19 on the psychological response by emphasizing depression-related issues. Similarly, the second and third themes focus on anxiety and stress among the healthcare workforce due to COVID-19.

TABLE 1.1 Participant's Description

SL. No.	Role as Caregiver	Age (Years)	Experience (Years)	Assigned Code
1.	Nurse	27	4	CG1
2.	Nurse	32	6	CG2
3.	Nurse	28	5	CG3
4.	Nurse	36	8	CG4
5.	Nurse	38	9	CG5
6.	Nurse	24	2	CG6
7.	Nurse	28	4	CG7
8.	Nurse	30	6	CG8
9.	Doctor	36	11	CG9
10.	Doctor	29	6	CG10
11.	Doctor	40	15	CG11
12.	Doctor	31	7	CG12
13.	Doctor	33	9	CG13
14.	Doctor	37	11	CG14
15.	Doctor	36	8	CG15

1.5.4   DATA ANALYSIS

Two individuals transcribed the recorded response from the participants in the English Language on permission from the participants. The RQDA (R package) for indexing code and categorizing the transcripts and establishing the relationship between the collected information using the simulation technique of Fruchterman-Reingold. The analysis presented in the next section includes the relationship between codes, categories, sub-themes, and themes. The first part of the discussion presents the analysis for the number of codes (119) and categories (94 refined to 49). The second part depicts the relationship between categories (49) and sub-themes (9). The relationship between sub-themes (9) and themes (4) is in the third part of the analysis. Table 1.2 presents the information in the codebook-

TABLE 1.2   Data Management

Particulars	Codebook
Open code	119
Initial categories	94
Refined categories	49
Sub-themes	9
Themes	4

1.6   RESULTS AND DISCUSSION

1.6.1   FACTORS AFFECTING PSYCHOLOGICAL NEED OF MEDICAL WORKFORCE-ANALYSIS BASED ON CATEGORIES AND SUB-THEMES DEVELOPED THROUGH ANALYSIS OF TRANSCRIPTS

To identify the factors affecting the psychological needs of the medical workforce, we imported the transcripts to RQDA. Figure 1.2 illustrates the relationship between codes and categories. The interpretation of Figure 1.2 includes Human needs linking with cognitive and esthetic needs, the satisfaction of needs for humans, the objective of human needs, and human need requirement. Physical health is another aspect of human needs, which states that needs merge within the body. Human needs can also be classified based on ego orientation and task orientation. In further analysis, we identified that the task orientation for the medical workforce during COVID-19 is a significant need for the concerned employees. The need for relatedness and means desire determine the ambitious human need as the next category. Furthermore, the need for self-esteem, lack of clarity, the fears of failures, criticism by family members of COVID-19 positive/ suspect cases, and high expectations from the government identifying as the codes from transcripts.

We have identified that for the medical workforce, autonomy, compe-tence, and relatedness in psychological needs determine broadly two categories, namely–in-born human needs and ambitious human needs. Basic human needs involve factors like the need for love, belongingness, psychological integrity, frustration, and these were in the responses from most participants. The base for basic human needs is on the fundamentals of psychological needs and understanding of a particular need. We found that the understanding of psychological needs was missing among the

Factors Affecting Psychological Needs of Medical Workforce

medical workforce during COVID-19. In addition to these factors, mental well-being, life experience, work-life balance, and hard work concerning psychological needs contributes to three distinct categories of basic human needs, life experience and well-being of humans. Furthermore, volatile needs are the needs that change due to the situation, and the medical workforce experiences such volatility needs to a significant extreme during pandemics such as COVID-19. Uncertainty in the future also determines psychological needs.

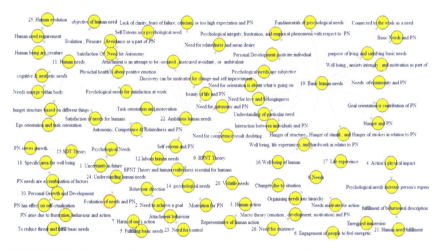

FIGURE 1.2 Relationship 1 using the Fruchterman Reingold layout algorithm.

1.6.2 FACTORS AFFECTING PSYCHOLOGICAL NEED OF MEDICAL WORKFORCE-ANALYSIS BASED ON CATEGORIES AND SUB-THEMES DEVELOPED THROUGH ANALYSIS OF TRANSCRIPTS

Figure 1.3 presents the relationship between categories and sub-themes. Factors like need preference, need to achieve goals, and human survival needs determine the perceived knowledge to manage the external physical situation. The other crucial factors in determining human needs include competence to manage psychological responses, capacity to manage psychological responses, confidence to manage psychological responses, confidence to manage the external physical situation, possible psychological responses to stress, and possible psychological responses to uncertainty in the medical workforce. BPNT theory, SDT theory, life experience, and human needs determine the perceived responsibility in managing the external physical situation. The factors like possible psychological responses to the uncertainty, competence to manage psychological response, perceived responsibility to manage the external physical situation, capacity to manage psychological response, confidence to manage psychological response, confidence to manage the external physical situation, probable psychological response to the stress and awareness and anticipation of probable psychological response also determines basic human needs. The factors include inborn human needs, ambitious human needs, basic human needs, and the preference for the mental well-being of the medical workforce determines competence to manage a psychological response.

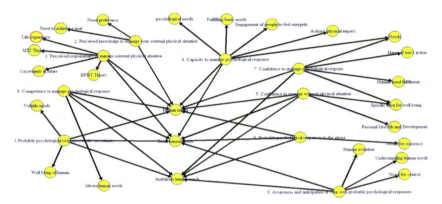

FIGURE 1.3 Relationship 2 using the Fruchterman Reingold layout algorithm.

Factors such as volatile needs, the well-being of humans, inborn human needs, basic human needs, and human needs describe possible psychological responses to uncertainty. We notice a robust relationship between the identified needs and the psychological needs that are highly interconnected. The capacity to manage psychological response includes psychological needs, fulfilling basic human needs, engaging people to feel energetic, action's physical impact, work-life balance, and emotional support during tough times.

The determination of confidence to manage psychological responses is by the support system available to the medical workforce, protective equipment, practical training of employees, good governance practices

Factors Affecting Psychological Needs of Medical Workforce   11

at the hospital, and government intervention for public awareness. The awareness and anticipation of possible psychological responses include human evolution, understanding human needs, the need for control, ambitious human needs, and basic human needs. Confidence to manage external physical situations includes specific mental well-being, personal growth, participation in the training programs, access to resources, interaction with a professional network, and support from regulatory authorities. Needs for existence, basic human needs, ambitious human needs, the experience of the medical workforce, and the support system for the medical workforce determine possible psychological responses to stress.

1.6.3   FACTORS AFFECTING PSYCHOLOGICAL NEED OF MEDICAL WORKFORCE-ANALYSIS BASED ON SUB-THEMES AND THEMES DEVELOPED THROUGH ANALYSIS OF TRANSCRIPTS

Figure 1.4 presents the relationship between sub-themes and themes. The anxiety among the medical workforce linking low competence to manage psychological responses, low perceived knowledge to manage the external physical situation, high perceived responsibility to manage the external physical situation, and the low capacity to manage the psychological response. Stress among the medical workforce is a cause of factors such as the possible psychological responses to uncertainty, low confidence to manage external and physical situations, low or missing awareness and anticipation of possible psychological responses, low capacity to manage psychological response, and low confidence to manage the psychological response. Confidence to manage the external psychological situation, in addition to determining stress, also determines depression. Possible psychological responses to stress, awareness of possible psychological responses, capacity to manage psychological responses, and competence to manage the psychological response causes depression. The majority of participants mentioned the lack of counseling facilities to share their psychological issues with the experts. The medical workforce also reported a lack of programs focusing on disaster situations during their training programs.

Table 1.3 presents the factors identified through semi-structured interviews for the psychological needs of medical caregivers.

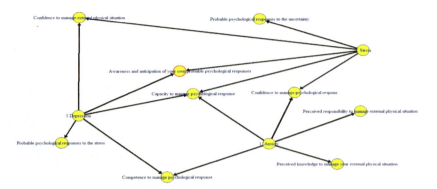

FIGURE 1.4 Relationship 3 using the Fruchterman Reingold layout algorithm.

TABLE 1.3 Factors Affecting the Psychological Needs of Caregivers During COVID-19

SL. No.	Factors	Participants Percentage
1.	High perceived responsibility to manage the external physical situation.	80
2.	Low perceived knowledge to manage their external physical situation	73
3.	Certain probable psychological responses to the uncertainty.	67
4.	Low perceived capacity to manage the psychological response.	60
5.	Low confidence to manage the external physical situation.	73
6.	Low perceived competence to manage the psychological response.	93
7.	Low awareness of their possible psychological responses to pandemics.	67
8.	Lack of protective equipment's and tools to treat COVID-19 positive patients.	80
9.	Long working hours in the hospitals due to COVID-19.	73
10.	Lower interaction with family members (social groups) during COVID-19.	73

The identified factors highlight the percentage of participants who suggested the factors. To our surprise, the majority of the respondent identified a standard set of psychological factors affecting the medical workforce. Around 93% of the participants considered low perceived competence to manage psychological responses as the viral factor affecting the psychological needs of healthcare workers. The factor affecting psychological needs links to the high-perceived responsibility to manage the external physical situation and lack protective equipment and tools to treat COVID-19 positive patients. The low perceived knowledge to manage the external physical situation, the low confidence to manage the external physical situation, the long working hours in the hospitals due to COVID-19, and the lower interaction with family members (social groups) during COVID-19 are core factors affecting psychological needs of the medical workforce. The early interventions of planners to design plans for managing the medical workforce's psychological stress are visible in the literature [19]. Proposed to focus on the rotation of workers as per their job profile, ensure fulfillment of basic psychological needs, pay attention to the vulnerable staff, create a culture of peer support, and a well-trained team for a psychological support medical workforce. The community-based model can help prepare [20] the community records, linking with the government database for dashboard through information and communication technology [21, 22]. The caregiver's well-being is key to the care recipient is who is, in turn, getting help from the caregiver in a nursing home at the recipient's home or the community center. The role of healthcare managers is critical to assess the psychological well-being of the medical workforce because the manager can take proactive steps to ensure the mental well-being of the employees.

1.7 CONCLUSION

The medical workforce experiences many changes in their psychological needs as an outcome of work pressure due to an increase in the caregiving to COVID-19 patients. This study shows that anxiety, stress, and depression among the medical workforce are increasing. This study identifies 10 psychological factors experience by the medical workforce. Anxiety, Depression, and Stress are the primary psychological needs increasing amongst the medical workforce under study. The psychological need assessment of caregivers is crucial for regulators who should focus on regular training and development of individuals at the front providing caregiving services. The government should ensure the medical workforce's counseling services and ensure a sufficient number of the trained medical workforce. Concisely, the medical workforce should possess full facilities, access to protective equipment, and the availability of online counseling facilities.

KEYWORDS

- **anxiety**
- **COVID-19**
- **depression**
- **medical workforce**
- **Middle East respiratory syndrome**
- **severe acute respiratory syndrome**
- **stress**

REFERENCES

1. Lu, W., Wang, H., Lin, Y., & Li, L., (2020). Psychological status of medical workforce during the COVID-19 pandemic: A cross-sectional study. *Psychiatry Research,* 112936.
2. Ahmed, W., Angel, N., Edson, J., Bibby, K., Bivins, A., O'Brien, J. W., & Mueller, J. F., (2020). First confirmed detection of SARS-CoV-2 in untreated wastewater in Australia: A proof of concept for the wastewater surveillance of COVID-19 in the community. *Science of the Total Environment, 728,* 138764.
3. Cole-King, A., & Dykes, L., (2020). *Well-Being for HCWs During COVID-19,* p. 8.
4. Greenberg, N., Docherty, M., Gnanapragasam, S., & Wessely, S., (2020). Managing mental health challenges faced by healthcare workers during the COVID-19 pandemic. *BMJ,* 368.
5. Jackson, D., Bradbury-Jones, C., Baptiste, D., Gelling, L., Morin, K., Neville, S., & Smith, G. D., (2020). Life in the pandemic: Some reflections on nursing in the context of COVID-19. *Journal of Clinical Nursing.*
6. Rajiah, S., Naidoo, J., Tirvassen, R., & Tyagi, R., (2021). A qualitative study on managing the transition practices of sending and receiving teachers in Mauritius. *International Journal of Management, 12*(3), 812–822. doi 10.34218/IJM.12.3.2021.079.
7. Calvo, R. A., Deterding, S., & Ryan, R. M., (2020). *Health Surveillance During a COVID-19 Pandemic,* (p. 369).
8. Gupta, S., Tyagi, R., Sharma, A., & Singh, H., (2021). Human wealthsurance: Analytical study on financial planning of community investors during COVID-19. *International Journal of Management, 12*(1), 1453–1473.
9. Tyagi, R., (2012). Meerut embroidery cluster: A case study. *South Asian Journal of Business and Management Cases, 1*(2), 185–202.
10. Tyagi, R. K., & Vasiljeviene, N., (2013). The case of CSR and irresponsible management practices. *Competitiveness Review, 23*(4, 5), 372–383.

11. Dai, M., Liu, D., Liu, M., Zhou, F., Li, G., Chen, Z., & Xiong, Y., (2020). Patients with cancer appear more vulnerable to SARS-COV-2: A multicenter study during the COVID-19 outbreak. *Cancer Discovery, 10*(6), 783–791.
12. Galea, S., Merchant, R. M., & Lurie, N., (2020). The mental health consequences of COVID-19 and physical distancing: The need for prevention and early intervention. *JAMA Internal Medicine, 180*(6), 817, 818.
13. Maben, J., & Bridges, J., (2020). COVID-19: Supporting nurses' psychological and mental health. *Journal of Clinical Nursing.*
14. Qiu, J., Shen, B., Zhao, M., Wang, Z., Xie, B., & Xu, Y., (2020). A nationwide survey of psychological distress among Chinese people in the COVID-19 epidemic: Implications and policy recommendations. *General Psychiatry, 33*(2).
15. Ripp, J., Peccoralo, L., & Charney, D., (2020). Attending to the emotional well-being of the health care workforce in a New York City health system during the COVID-19 pandemic. *Academic Medicine.*
16. Pereira-Sanchez, V., Adiukwu, F., El Hayek, S., Bytyçi, D. G., Gonzalez-Diaz, J. M., Kundadak, G. K., & Ransing, R., (2020). COVID-19 effect on mental health: Patients and workforce. *The Lancet Psychiatry, 7*(6), e29, e30.
17. Wang, C., Pan, R., Wan, X., Tan, Y., Xu, L., McIntyre, R. S., & Ho, C., (2020). A longitudinal study on the mental health of the general population during the COVID-19 epidemic in China. *Brain, Behavior, and Immunity.*
18. Zhang, W. R., Wang, K., Yin, L., Zhao, W. F., Xue, Q., Peng, M., & Chang, H., (2020). Mental health and psychosocial problems of medical health workers during the COVID-19 epidemic in China. *Psychotherapy and Psychosomatics, 89*(4), 242–250.
19. Billings, J., Kember, T., Greene, T., Grey, N., El-Leithy, S., Lee, D., & Bloomfield, M., (2020). Guidance for planners of the psychological response to the stress experienced by hospital staff associated with COVID: Early interventions. *Occupational Medicine*, kqaa098.
20. Tyagi, R., Vishwakarma, S., Yadav, S. S., & Stanislavovich, T. A., (2020). Community self-help projects. In: Leal, F. W., Azul, A., Brandli, L., Lange, S. A., Özuyar, P., & Wall, T., (eds.), *No Poverty: Encyclopedia of the UN Sustainable Development Goals.* Springer, Cham.
21. Tyagi, R., Vishwakarma, S., Alexandrovich, Z. S., & Mohammed, S., (2020). ICT skills for sustainable development goal 4. In: Leal, F. W., Azul, A., Brandli, L., Lange, S. A., Özuyar, P., & Wall, T., (eds.), *Quality Education: Encyclopedia of the UN Sustainable Development Goals.* Springer, Cham.
22. Tyagi, R., Vishwakarma, S., Singh, K. K., & Syan, C., (2020). Low-cost energy conservation measures and behavioral change for sustainable energy goal. In: Leal, F. W., Azul, A. M., Brandli, L., Lange, S. A., & Wall, T., (eds.), *Affordable and Clean Energy: Encyclopedia of the UN Sustainable Development Goals.* Springer, Cham.
23. (2020). *WHO Coronavirus [COVID-19] Dashboard.* Available at: https://COVID19. who.int (accessed on 5 July 2022).

CHAPTER 2

Impact of COVID-19 on Employees' Emotional Health and the Role of Leadership

MANSI BABBAR[1] and ISHITA KHANNA[2]

[1]*Department of Commerce, Shaheed Bhagat Singh College, University of Delhi, Delhi–110017, India, E-mail: mansibabbar21@gmail.com*

[2]*University of Delhi, Delhi–110017, India*

ABSTRACT

While more lethal strains of the virus have wrecked and ruined mankind in recent decades, none have attained the level of global superstardom that the novel coronavirus (nCoV) has. As the humanitarian disaster from the coronavirus pandemic unfolds, the economies are on the verge of unprecedented economic catastrophe, forced alterations are happening in the working climate, and businesses are grappling with uncertainty about the future, which is adversely impacting both employees and organizations. Employees are not only physically affected by the infection, but also psychologically, as they suffer a slew of strong emotional fluctuations. In this regard, the present study intends to synthesize and analyze the existent literature on the subject in order to better understand the emotional consequences of pandemic on employees and businesses at large. A thorough evaluation of literature propounds that the precarious crisis resulted in various negative emotional outcomes such as stress, fear, anxiety, tension, burnout, and depression. The study also asserts that detrimental emotional

---

Emotional Intelligence for Leadership Effectiveness: Management Opportunities and Challenges During Times of Crisis. Mubashir Majid Baba, Chitra Krishnan, & Fatma Nasser Al-Harthy (Eds.)
© 2023 Apple Academic Press, Inc. Co-published with CRC Press (Taylor & Francis)

repercussions can have grave ramifications for organizations, underscoring the importance of leadership in dealing with emotional crisis. Against this backdrop, suggestions are presented that may assist employees and organizations in enduring and sustaining any future crisis circumstances.

2.1 INTRODUCTION

On March 11, 2020, the World Health Organization (WHO) declared the coronavirus disease 2019 (COVID-19) a pandemic. It is an infectious disease induced by the novel Coronavirus (nCoV) which has not been previously identified or studied in humans. It has paralyzed industries, debilitated economies, and has confined entire mankind to their homes. The government-mandated norms such as quarantine, social distancing, and lockdowns resulted in the closure of physical premises and caused dramatic changes in the working lives where work from home policies were introduced for most employees around the globe. Hence, the virus not only impacted employees physically, but also psychologically [1, 2], due to abrupt changes in their working and personal lives wherein employees experienced a plethora of emotional fluctuations.

It was witnessed that the extremely high transmission rate of the virus along with precautionary confinement resulted in various emotions such as fear, anxiety, anger, sadness, loneliness, and frustration [3–5], which caused immense emotional turmoil among people. Further, disruption in personal and professional lives, job instability, financial setbacks, and continuous widespread dissemination of negative news through social media contributed to emotional swings where employees experienced tremendous stress, tension, anxiety, and depression [6–11].

In view of the above, the current research aims to highlight the impediments and emotional interactions that employees experienced during the pandemic. The study also aims to investigate the pandemic-induced factors that led to extreme emotional consequences. It has been noted that research on employees in the sense of pandemic, is still in its infancy, with most studies dealing only tangentially with the subject. Hence, the present research makes a significant and elaborate contribution by synthesizing, interpreting, and presenting the existing literature on pandemic-provoked employee emotional experiences and reactions. It is also important to note that in such precarious times, the role of leadership in the organizations becomes crucial as they have to keep the employees motivated even amidst the crisis. To curb the adverse impact of employees' vulnerable emotional states on organizational outcomes, it is imperative that leaders come forward and lead the employees in overcoming their emotional fluctuations. In this regard, the main objective of this study is to conduct a thorough analysis of the existing literature, and it aims to answer the following two research questions as diagrammatically presented in Figure 2.1.

> ➢ **RQ1:** How has COVID-19 pandemic impacted employees' emotional health?
> ➢ **RQ2:** What role does leadership play in managing such emotional crisis situations?

The understanding of emotional impact of pandemic on human resources (HRs) will enable the organizations to accurately assess and effectively evaluate the overall impact of crisis on their business operations. This will also equip leaders and managers to take preventive and corrective actions to restore employee mental and psychological well-being, which not only affects them individually but also has an adverse impact on organizations and economies at large.

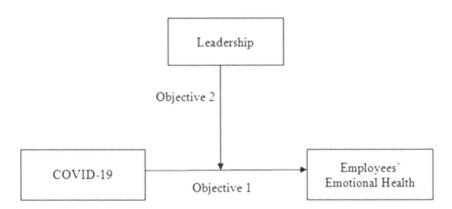

FIGURE 2.1 Objectives of the study.

2.2 METHOD

The extensive review of literature was performed between January and April 2021 based on the guidelines of Gough, Oliver, and Thomas [12].

The research articles were searched in renowned databases such as Emerald, JSTOR, Springer, SAGE, Wiley, Scopus, and Science Direct. Along with this, manual search in reputed management journals and Google scholar was also done to include all relevant articles in the study. Further, considering the topical nature of the subject, few reports from international organizations and social media also guided the identification of pandemic-induced emotional outcomes for the development of the proposed conceptual framework. The key strings used for searching the databases were "COVID-19 and emotions," "COVID-19 or coronavirus or pandemic and employees," "COVID-19 or pandemic and organizations," "COVID-19 or pandemic and workplace," "crisis and leadership," "COVID-19 or pandemic and leadership," and "emotions and leadership." This resulted in a total of 96 articles. Few articles which marginally dealt with the topic were excluded from analysis and hence, a total of 87 articles were considered relevant for the present study. Articles published in English language were only considered and no other restriction was made.

2.3 EMOTIONAL IMPACT OF PANDEMIC ON EMPLOYEES

The pandemic has impacted millions of people not only physically but also emotionally where several academics have stated that during times of crisis, people experience fluctuating emotions and extreme emotional states [3, 5, 13]. It was observed that the very high transmission rate of coronavirus along with social constraints resulted in various emotions such as fear, tension, anxiety, and hopelessness which consequently resulted in intense emotional states of stress among people [3, 4, 14, 15]. Studies also suggest that anxiety, tension, and loneliness, as experienced during pandemic, disturb the psychological health, as they are strongly linked to depression [14, 16] which further escalates sadness and negative emotions [17]. Further, it is advocated that the social isolation as mandated by the government, where people experienced immense loneliness, is also strongly associated with intense emotional experiences of anxiety and stress [13, 14, 16]. Such government-mandated norms also caused abrupt lifestyle changes wherein distancing, confinement, and loss of usual routine caused boredom and frustration among people [2, 4], which further escalated and intensified their emotional experiences. It is implicated that such encounter with negative emotions, an increase in virus-infected people, and rising mortality tolls is likely to lead to emotional disorders such as anxiety, personality changes, depression, bipolar, and cognitive problems [3].

Further, job and workspaces which are important aspects of people's lives, underwent disruption as the COVID-19 pandemic wreaked havoc and raised unparalleled obstacles in the professional lives of employees. Due to preventive measures such as distancing and lockdowns, most organizations introduced remote work from home policies for their employees where it was witnessed that maintaining work-life balance became a serious challenge as work started interfering with family commitments and vice versa [18, 19]. In such perilous times, the separation of personal and professional life became a major challenge for employees and literature suggests that such situations of dilemma can lead to increased frustration, stress, emotional exhaustion, and burnout among employees [10]. The emotional experiences were even more severe for those who were working outside homes during lockdowns for essential services such as pharmacies, clinics, hospitals, and grocery stores, where employees were constantly exposed to the risk of infection. It was reported in the news that at some places, such employees were even stigmatized as possible carriers of coronavirus, rather than being appreciated for their services in dangerous times, which further exacerbated burnout among them [20]. The employee burnout, which has emotional exhaustion as one of its dimensions [21], also enhances due to other factors such as less social contact with family and friends, insecurity about job and finances, and recession caused by economic slowdown [6, 9, 11, 22], all of which have been encountered by employees during the unprecedented pandemic.

The literature also suggests that social media plays a huge role in influencing the emotional experiences of people. Kramer et al. [23] advocates that emotions can be conveyed to others not only physically, but also virtually through social media which has now become an indispensable part of life. It is witnessed that the usage of digital social media increased during crisis time as it was considered convenient to obtain real-time notifications and coronavirus cases updates. However, the constant media exposure, negative news, and fake news from unreliable sources caused anxiety and stress among people [24, 25]. Findings of research indicate that news headline connotations have very high emotion scores that induce negativity [3]. The quick and huge amount of information spread of COVID-19 pandemic on news and media made WHO term it as "infodemic" [3] where the literature highlights a strong correlation between excessive use of social media and depressive symptoms among people. For example, Shacham et al. [26] advocates that employees' psychological health is greatly threatened by overly constant exposure to negative news regarding the pandemic, which resultantly increases emotional strains and the level of depression among them.

Such emotional responses were seen to further amplify in employees as they encountered news of pandemic-induced layoffs, furloughs, recession, and economic downturn [7, 25]. It was observed that as the economic activity completely stagnated, many firms were unable to bear the expense of their employees. This resulted in widespread layoffs and unemployment which not only caused financial distress, but also emotional distress along with amplified emotional states of depression [22, 27, 28]. Hamouche [29] also advocates that job insecurity and financial losses during the COVID-19 pandemic may be considered as strong and long-lasting stressors that deeply impact the emotional and mental health of employees. Wilson [30] also suggests that the rates of anxiety and depression during the pandemic may have accelerated due to concerns about employment status and financial instability. Insecurity about job and finances is a stressful experience, which causes distress and results in negative emotions and feelings [30]. Losing a job is not the only concern, but difficulty in getting new employment in crisis time is also an equally disturbing concern, which instills hopelessness and causes worry, which consequently results in poor mental health [30].

The employees also experienced emotional dissonance, which is defined as a deviation and inconsistency between the emotions that are experienced and expressed [31, 32]. Various researchers propagate that during COVID-19 crisis, employees endured a wide range of emotions due to disruptions in their personal and professional lives, such as loss of close relatives, infected dear ones, inappropriate workspaces, extended work hours, and continuous fear of infection [33, 34]. However, despite such emotional disturbances and disruptions, employees continued to work, which caused emotional and cognitive dissonance that further elevated emotional exhaustion and burnout among employees [1]. The variation between displayed and authentic emotions is a matter of grave concern as it not only impacts employees, but also gets reflected in organizational performance [1].

Further, emotion contagion, a social psychology concept, also gained limelight in the context of COVID-19 pandemic. It refers to the well-established notion that one person's emotions and affective behavior can be influenced by that of others. Considering the devastating pandemic crisis, studies suggest that it is possible that those who are more sensitive to others' emotions will be more emotionally and psychologically disturbed during the pandemic [5]. The findings of a research study reveal that individuals with a higher susceptibility to emotion contagion, showed higher levels of anxiety, stress, depression, and OCD (obsessive-compulsive disorder) symptoms during COVID-19 [5]. Therefore, for survival of the organizations during pandemic, it is necessary to understand the variety and intensity of emotions experienced by the employees. This emphasizes that managing the emotional fluctuations and leading the employees through the crisis is crucial for the smooth functioning of organizations amidst crisis. Hence, the value of a leader's intervention in parlous times cannot be underestimated.

2.4 THE ROLE OF LEADERSHIP

The unprecedented crisis forced mankind to embrace change and has taught not only corporate leaders, but the entire world, a variety of vital lessons. The COVID-19 epidemic has had a detrimental impact on the emotional health of employees; thus, it is important to understand the role of leadership in addressing these situations. It is already known that the achievement criteria for crisis management begin with "leadership" [35] and its importance is well-established in literature. Employee performance can be affected by the emotional climate created by a leader, according to prior research [36]. According to the literature, there are many leadership skills that are useful during times of crisis. Integrity, intelligence, vision, charisma, authenticity, communication, influence, emotion management, participatory decision-making, self-awareness, and effective sense-making skills are some of them [37–39]. As a result, effective leadership is essential to respond to market changes, creatively handle challenges, steer the business through rough phase, secure its survival, and maintain high performance during times of crisis [40, 41]. In this light, effective leadership is regarded as the bedrock of organizational growth and performance, and may be determined by factors such as emotional stability, self-efficacy, emotional intelligence (EI), self-monitoring, conscientiousness, and extraversion [42, 43].

Transformational leadership (TL) is commonly acknowledged as the most effective style of leadership [44]. Bass and Avolio [45] described TL as the leadership in which the leader arouses the interest of co-workers and followers to look at their work with new perspectives. They claimed that TL had four dimensions, which they call the "Four I's" viz. idealized influence, intellectual stimulation, inspirational motivation, and individualized consideration. It is believed that the TL style can be helpful in leading the employees through emotional turmoil as literature suggests that leaders have an indirect impact on the psychological health of their employees [46]. Shamir et al. [47] suggest that transformational leaders might assist employees to deal with emotional regulation in more efficient and less psychologically taxing ways. It is stressed that both negative and positive emotions may be regulated, and both emotion expression and experience may be addressed [48]. Incorporating the theory of conservation of resources (COR) given by Hobfoll [49] as a theoretical framework, Walsh, Dupre, and Arnold [50] argue that transformational leaders influence employee well-being. The COR theory posits that individuals exhibit bias in overweighing resource loss and underweighting resource gain. It advocates that stress occurs when the valued resources of individuals such as family, health, well-being, self-esteem, and meaning in life are threatened with loss. Such resource loss negatively impacts the employee well-being [49], which emphasizes the importance of leadership intervention. The leaders can invest their own resources (e.g., social support, self-efficacy, and role clarity) in the production of resources for followers who nurture and expand these resources as it provides meaning to them and elevates their well-being. There is a strong correlation between TL and employee wellness, according to recent studies. Findings demonstrate that transformative leadership is favorably connected with employee psychological health, encompassing both good and negative consequences. (e.g., optimism and confidence) and negative (e.g., anxiety and stress) conceptualizations of health [51]. Prior research suggests that transformational leaders not only improve employees' psychological health, but also foster their commitment to the organization and encourage them to improve their performance. In the context of COVID-19's crisis, the importance of TL is underscored [52, 53]. As a result, in a pandemic-affected organization with shifting demands and falling performance, the TL style may be the most appropriate, where leaders encourage and guide employees through the changes.

Research experts are increasingly embracing the idea of EI as a potential attribute of good leadership. EI is defined as the ability to understand and control one's own and others' feelings [54]. Since the early 1990s, researchers have been studying the impact of EI at work. Particularly among leaders [55] and have propagated that EI can greatly influence a person's leadership skills [56, 57]. Numerous scholars have advocated that a transformational leader manifests qualities including self-awareness, empathy, self-confidence, and motivation [58–60], whereas Goleman [61] described all of the above-mentioned attributes as subcomponents of EI. EI and TL styles have several other features in common [62]. It is believed that leaders with high EI are better able to handle stressful situations and disagreements by effectively managing their own and others' emotions. Hence, EI is employed as a measure of leadership effectiveness [63–65].

Further, considering the present times dominated by unpredictability, uncertainty, and dynamism, leading requires efficient emotion regulation (ER), which is defined as the process of influencing one's emotions and modifying one's emotional manifestations [48, 65]. ER has been viewed as a crucial competency related to effective leadership [36, 66, 67]. It is believed that leaders with an intentional approach to ER are effectively able to solve problems and they avoid rigidity in making decisions [68], thus enhancing comfort and openness to challenges. They manage their own as well as their employees' emotions, and by doing so, they significantly impact their employees' performance [69] and preserve employees' positive emotions [70]. ER strategies such as cognitive reappraisal and situation modification appear to be functional strategies for dealing with emotion-laden events at work [71]. Although exercising ER is a contentious topic, with experts supporting both advantages and drawbacks, it is critical in the context of a crisis as employees are exposed to a large flood of highly contagious negative emotions. Thus, to eliminate the adverse chain effects of extreme negative contagion, leaders may adopt ER, which positively influences job performance [65], and consequently contributes towards the organization as a whole. However, in recent years, it has been suggested that leaders are also expected to display their positive emotions as a part of their job. Research in this field suggests that leaders who display emotions may affect their subordinates through emotional contagion [72, 73], wherein subordinates internalize and experience perceived leader emotions [74] which instills positivity in them and results in favorable outcomes.

Thus, it can be concluded that leadership plays a crucial role in managing emotional crisis situations such as that caused by the COVID-19 pandemic. As a result, crises provide a unique chance for leaders to showcase their ability to lead effectively [75]. Research shows that there is a significant link between the conduct of leadership, the health of individuals and the working environment, and also between leadership styles and the results of healthy and productive organizations [76]. Therefore, the ability to understand and manage moods, styles, feelings, and emotions in oneself and others is critical to effective leadership in organizations [77] and is especially important in pandemic-stricken perilous times.

2.5 DISCUSSION

The COVID-19 pandemic has caused both financial and non-financial disruptions to businesses and organizations. Though financial disruptions are easily gauged and measured, non-financial disruptions in terms of adverse mental, emotional, and psychological health of HRs of organization largely remain outside the ambit of disruption measurement. Also, it is important to note that the two are not mutually exclusive, i.e., financial disruptions may set grounds for simultaneous non-financial disruptions. Previous research also advocates those disasters which cause large financial problems for individuals tend to be linked to high levels of severe and long-lasting psychological effects [29, 78]. The emotional fluctuations induced by the pandemic have proved detrimental to both existing and former employees. On one hand, the widespread furloughs and layoffs have caused tremendous financial loss to employees and on the other hand, it instilled fear and job insecurity amongst the existing workforce. The literature also highlights that curtailment in staffing levels tends to reduce the level of organizational commitment and job involvement, and elevates stress levels among existing employees [79]. This deeply impacts their emotional health and thus, calls for the immediate attention of leaders to lead the employees through the emotional crisis. The leadership has a significant role and responsibility to ensure the emotional well-being of employees and the growth of the organization. Therefore, emotionally intelligent leadership is seen as a probable solution to combat the fierce emotional fluctuations experienced by HRs during the time of COVID-19 pandemic.

Impact of COVID-19 on Employees' Emotional Health 27

Further, research shows that people, who are engulfed with negative emotions and are depressed, are less inclined to embrace COVID-19 preventative measures such as quarantines, mass gathering cancellations, and business closures [80]. This is because the government-mandated closure regulations and work-from-home guidelines caused anxiety, fear, burnout, and depression, as aforementioned, which made people uncooperative and unconcerned about prevention measures, which further escalated the spread of the coronavirus. Hence, the role of leadership is not only limited to employees and organizations, but extends beyond and indirectly contributes to curbing the spread of the virus by keeping employees emotionally healthy, vigilant, and aware of the pandemic protocols that need to be sincerely followed despite turmoil.

It is also worth noting that the gap between high-skilled and low-skilled jobs may widen due to the economic downturn brought by the pandemic. The organizational investment in employees employed in high-skilled occupations is assumed to rise since they help organizations maintain their productivity even in adverse situations, such as those posed by disease outbreaks [81]. The low-skilled occupations, on the other hand, are more likely to be overlooked by management because they do not contribute significantly to organizational productivity. Further, with prolonged lockdowns and complete movement restrictions coupled with grave future uncertainty, the demand for products reduced and investments decreased, which resulted in huge job losses and business closures. In this light, the strategic role of leadership needs to be highlighted where the leaders strive to not only manage the employees, but also undertake responsibility for the survival of the organization and devise appropriate strategies considering prevailing market situations.

2.6 RECOMMENDATIONS

COVID-19 pandemic has elevated the significance of emotional and mental health to the forefront of the corporate agenda, as it intensified the employees' emotions of fear, anxiety, and depression which are detrimental to the survival and growth of organizations. Considering the tempestuous times, it is imperative that corporates introduce employee assistance policies and practices to safeguard their employees' physical as well as emotional well-being. Regular counseling and therapy sessions, crisis awareness campaigns, and psychological resources, including social support and feedback through regular virtual meetings with employees are all part of this [7, 82]. In the wake of the pandemic crisis, it was witnessed that many corporates throughout the world, including Facebook, HCL (Hindustan Computers Limited), Accenture, Infosys, Godrej, Google, Wipro, and Cognizant incorporated multiple support action plans to deliver value for their employees [83]. For example, in the wake of the coronavirus outbreak, Facebook announced new paid-time-off initiatives which allowed employees to take up to a month off from work to look after their sick relatives [84]. Companies like Infosys established COVID-19 Care Centers (CCC) for employees and their families, and Cognizant came forward and provided immunizations to all of its employees and their family members, as well as medical assistance for COVID-19 positive associates and their dependents, paid time off for recovery, and emergency financial assistance to junior staff in urgent situations to help pay medical expenses [85].

Furthermore, the government intervention is important to initiate financial security measures in order to decrease the prevalence of employee mental disorders during the disease outbreak. For instance, governments in nations such as the United Kingdom, Spain, and France, have devised emergency packages that involve granting loans and guarantees to businesses to reduce the negative economic effects of the COVID-19 outbreak, as well as direct reimbursements to employees to alleviate financial instability [86]. On the other hand, leaders may adopt an emotionally intelligent way and an open leadership style that encourages workers to work cheerfully, decreases work stress and corresponding negative behavior, and supports corporate growth [87]. Thus, to deal with the pandemic's spillover effect, all stakeholders, including leaders, policymakers, governments, and corporations, should work together to implement policies and practices at work that will reduce productivity losses and improve employees' psychological health.

2.7 CONCLUSION

The tremendous devastation caused by the COVID-19 pandemic has brought trade and businesses to a screeching halt. It has swamped healthcare facilities, crippled economies, wreaked havoc on communities, and

Impact of COVID-19 on Employees' Emotional Health

forced millions of people to stay in their homes. It is witnessed that the ambiguity and the uncertainty connected with the outbreak of coronavirus disease pushed employees into a tailspin and intensely impacted their psychological health. In this context, the present study underlines the impact of the COVID-19 catastrophe on employees' emotional health and accentuates the role of leadership in managing emotional crisis situations. However, this has not only impacted employees but has also uprooted big corporations and caused an economic stalemate with no clear conclusion in sight. Thus, such perilous times and challenges offered by the pandemic call for transformational leaders to practice EI and ER strategies that assist employees and organizations sustain the crippling crisis. Hence, one of the hard-learned takeaways from the COVID-19 crisis is that we need to revitalize the health infrastructure, engage in more discussions about mental health issues in open and public forums, and understand the employee emotions and repercussions to lead them successfully through the crisis.

KEYWORDS

- **COVID-19**
- **crisis**
- **emotion regulation**
- **leadership**
- **mental health**
- **novel coronavirus**
- **obsessive-compulsive disorder**
- **pandemic**
- **World Health Organization**

REFERENCES

1. Baba, M. M., (2021). Emotional dissonance and exhaustion among library professionals during COVID-19. *Library Philosophy and Practice (E-Journal),* 4558. Retrieved from: https://digitalcommons.unl.edu/libphilprac/4558 (accessed on 5 July 2022).

2. Restubog, S. L. D., Ocampo, A. C. G., & Wang, L., (2020). Taking control amidst the chaos: Emotion regulation during the COVID-19 pandemic. *Journal of Vocational Behavior, 119*, 103440.
3. Aslam, F., Awan, T. M., Syed, J. H., Kashif, A., & Parveen, M., (2020). Sentiments and emotions evoked by news headlines of coronavirus disease (COVID-19) outbreak. *Humanities and Social Sciences Communications, 7*(1), 1–9.
4. Brooks, S. K., Webster, R. K., Smith, L. E., Woodland, L., Wessely, S., Greenberg, N., & Rubin, G. J., (2020). The psychological impact of quarantine and how to reduce it: Rapid review of the evidence. *The Lancet, 395*(10227), 912–920.
5. Wheaton, M. G., Prikhidko, A., & Messner, G., (2020). Is fear of COVID-19 contagious? The effects of emotion contagion and social media use on anxiety in response to the coronavirus pandemic. *Frontiers in Psychology, 11*, 567379.
6. Lindström, C., Åman, J., & Norberg, A. L., (2011). Parental burnout in relation to sociodemographic, psychosocial and personality factors as well as disease duration and glycaemic control in children with type 1 diabetes mellitus. *Acta Paediatrica., 100*(7), 1011–1017.
7. Majeed, M., Irshad, M., Fatima, T., Khan, J., & Hassan, M. M., (2020). Relationship between problematic social media usage and employee depression: A moderated mediation model of mindfulness and fear of COVID-19. *Frontiers in Psychology, 11*, 557987.
8. Maslach, C., Schaufeli, W. B., & Leiter, M. P., (2001). Job burnout. *Annual Review of Psychology, 52*(1), 397–422.
9. Parkes, A., Sweeting, H., & Wight, D., (2015). Parenting stress and parent support among mothers with high and low education. *Journal of Family Psychology, 29*(6), 907.
10. Sonnentag, S., Kuttler, I., & Fritz, C., (2010). Job stressors, emotional exhaustion, and need for recovery: A multi-source study on the benefits of psychological detachment. *Journal of Vocational Behavior, 76*(3), 355–365.
11. Sorkkila, M., & Aunola, K., (2020). Risk factors for parental burnout among Finnish parents: The role of socially prescribed perfectionism. *Journal of Child and Family Studies, 29*(3), 648–659.
12. Gough, D., Oliver, S., & Thomas, J., (2017). *An Introduction to Systematic Reviews.* London: Sage Publications Ltd.
13. Meléndez, J. C., Satorres, E., Reyes-Olmedo, M., Delhom, I., Real, E., & Lora, Y., (2020). Emotion recognition changes in a confinement situation due to COVID-19. *Journal of Environmental Psychology, 72*, 101518.
14. Baba, M. M., (2020). Navigating COVID-19 with emotional intelligence. *International Journal of Social Psychiatry, 66*(8), 810–820.
15. Joiner, T. E., Wingate, L. R., & Otamendi, A., (2005). An interpersonal addendum to the hopelessness theory of depression: Hopelessness as a stress and depression generator. *Journal of Social and Clinical Psychology, 24*(5), 649–664.
16. Holmes, E. A., O'Connor, R. C., Perry, V. H., Tracey, I., Wessely, S., Arseneault, L., & Bullmore, E., (2020). Multidisciplinary research priorities for the COVID-19 pandemic: A call for action for mental health science. *The Lancet Psychiatry, 7*(6), 547–560.

17. Cummins, N., Scherer, S., Krajewski, J., Schnieder, S., Epps, J., & Quatieri, T. F., (2015). A review of depression and suicide risk assessment using speech analysis. *Speech Communication, 71,* 10–49.
18. (2020). *COVID-19 Crisis: These Companies Allow Employees to Work From Home 'Forever'.* Business Today. https://www.businesstoday.in/current/corporate/covid-19-crisis-these-companies-allow employees-to-work-from-home-forever/story/418205.html (accessed on 5 July 2022).
19. Ramanujam, K., (2020). *Empowering the Permanently Remote Workforce.* Tata Consultancy Services. Retrieved from: https://www.tcs.com/blogs/empowering-the-permanently-remote-workforce (accessed on 5 July 2022).
20. Semple, K., (2020). *'Afraid to be a Nurse': Health Workers Under Attack.* The New York Times. Retrieved from: https://www.nytimes.com/2020/04/27/world/americas/coronavirus-health-workers-attacked.html (accessed on 5 July 2022).
21. Maslach, C., Jackson, S. E., Leiter, M. P., Schaufeli, W. B., & Schwab, R. L., (1986). *Maslach Burnout Inventory* (Vol. 21, pp. 3463–3464). Palo Alto, CA: Consulting psychologists press.
22. Griffith, A. K., (2020). Parental burnout and child maltreatment during the COVID-19 pandemic. *Journal of Family Violence.* https://doi.org/10.1007/s10896-020-00172-2.
23. Kramer, A. D., Guillory, J. E., & Hancock, J. T., (2014). Experimental evidence of massive-scale emotional contagion through social networks. *Proceedings of the National Academy of Sciences, 111*(24), 8788–8790.
24. Gao, J., Zheng, P., Jia, Y., Chen, H., Mao, Y., Chen, S., Wang, Y., Fu, H., & Dai, J., (2020). Mental health problems and social media exposure during COVID-19 outbreak. *Plos One, 15*(4), e0231924. https://doi.org/10.1371/journal.pone.0231924.
25. Garfin, D. R., Silver, R. C., & Holman, E. A., (2020). The novel coronavirus (COVID-2019) outbreak: Amplification of public health consequences by media exposure. *Health Psychology, 39*(5), 355–357.
26. Shacham, M., Hamama-Raz, Y., Kolerman, R., Mijiritsky, O., Ben-Ezra, M., & Mijiritsky, E., (2020). COVID-19 factors and psychological factors associated with elevated psychological distress among dentists and dental hygienists in Israel. *International Journal of Environmental Research and Public Health, 17*(8), 2900.
27. Brand, J. E., Levy, B. R., & Gallo, W. T., (2008). Effects of layoffs and plant closings on subsequent depression among older workers. *Research on Aging, 30*(6), 701–721.
28. Burgard, S. A., Brand, J. E., & House, J. S., (2007). Toward a better estimation of the effect of job loss on health. *Journal of Health and Social Behavior, 48*(4), 369–384.
29. Hamouche, S., (2020). COVID-19 and employees' mental health: Stressors, moderators and agenda for organizational actions. *Emerald Open Research, 2,* 15.
30. Wilson, J. M., Lee, J., Fitzgerald, H. N., Oosterhoff, B., Sevi, B., & Shook, N. J., (2020). Job insecurity and financial concern during the COVID-19 pandemic are associated with worse mental health. *Journal of Occupational and Environmental Medicine, 62*(9), 686–691.
31. Holman, D., Martinez-Inigo, D., & Totterdell, P., (2008). Emotional labor, well-being, and performance. In Cartwright, S., & Cooper, C., (eds.), *The Oxford Handbook of Organizational Well-Being* (pp. 331–355). Oxford University Press.

32. Zapf, D., (2002). Emotion work and psychological well-being: A review of the literature and some conceptual considerations. *Human Resource Management Review, 12*(2), 237–268.

33. Hossain, M. M., Sultana, A., & Purohit, N., (2020). *Mental Health Outcomes of Quarantine and Isolation for Infection Prevention: A Systematic Umbrella Review of the Global Evidence.* Retrieved from: https://ssrn.com/abstract=3561265 (accessed on 5 July 2022).

34. Sasangohar, F., Jones, S. L., Masud, F. N., Vahidy, F. S., & Kash, B. A., (2020). Provider burnout and fatigue during the COVID-19 pandemic: Lessons learned from a high-volume intensive care unit. *Anesthesia and Analgesia, 131*(1), 106–111. doi: 10.1213/ANE.0000000000004866.

35. Shelton, K., (1997). *Beyond So-Called Leadership* (p. 12). Istanbul: Rota Publishing.

36. Humphrey, R. H., (2002). The many faces of emotional leadership. *The Leadership Quarterly, 13*(5), 493–504.

37. Bolman, L. G., & Deal, T. E., (1997). *Reframing Organizations.* Jossey-Bass, San Francisco, CA.

38. Burnett, J., (2002). *Managing Business Crises: From Anticipation to Implementation.* Greenwood Publishing Group.

39. Furst, S. A., & Reeves, M., (2008). Queens of the hill: Creative destruction and the emergence of executive leadership of women. *The Leadership Quarterly., 19*(3), 372–384.

40. Vardiman, P. D., Houghton, J. D., & Jinkerson, D. L., (2006). Environmental leadership development: Toward a contextual model of leader selection and effectiveness. *Leadership & Organization Development Journal, 27*(2), 93–105.

41. Mumford, M. D., Friedrich, T. L., Caughron, J. J., & Byrne, C. L., (2007). Leader cognition in real-world settings: How do leaders think about crises?. *The Leadership Quarterly, 18*(6), 515–543.

42. McCauley, C. D., & Van, V. E., (2004).*The Center for Creative Leadership Handbook of Leadership Development* (2nd edn.). Jossey-Bass: San Francisco.

43. Kim, S., (2007). Learning goal orientation, formal mentoring, and leadership competence in HRD. *Journal of European Industrial Training, 31*(3), 181–194.

44. Baba, M. M., (2019). Transformational leadership and personal demographic profile in the education system of India. *Global Business Review.* https://doi.org/10.1177/0972150919884200.

45. Bass, B. M., & Avolio, B. J., (1994). *Improving Organizational Effectiveness Through Transformational Leadership.* SAGE Publications.

46. Krishnan, C., Goel, R., Singh, G., Bajpai, C., & Malik, P., (2017). Emotional intelligence: A study on academic professionals. *Pertanika Journal of Social Sciences & Humanities.*

47. Sivanathan, N., Arnold, K. A., Turner, N., & Barling, J., (2004). Leading well: Transformational leadership and well-being. In Linley, P. A., & Joseph, S., (eds.), *Positive Psychology in Practice* (pp. 241–255). Hoboken, NJ: John Wiley & Sons Inc.

48. Shamir, B., House, R. J., & Arthur, M. B., (1993). The motivational effects of charismatic leadership: A self-concept based theory. *Organization Science, 4*(4), 577–594.

49. Gross, J. J., (1998). The emerging field of emotion regulation: An integrative review. *Review of General Psychology, 2*(3), 271–299.
50. Hobfoll, S. E., (1989). Conservation of resources: A new attempt at conceptualizing stress. *American Psychologist, 44*(3), 513.
51. Walsh, M., Dupré, K., & Arnold, K. A., (2014). Processes through which transformational leaders affect employee psychological health. *German Journal of Human Resource Management, 28*(1, 2), 162–172.
52. Kelloway, E. K., & Barling, J., (2010). Leadership development as an intervention in occupational health psychology. *Work & Stress, 24*(3), 260–279.
53. Bass, B. M., (1998). *Transformational leadership: Industrial, Military, and Educational Impact.* Lawrence Erlbaum Associates Publishers.
54. Bass, B. M., & Riggio, R. E., (2006). *Transformational Leadership* (2nd edn.). Psychology Press. Retrieved from: https://www.routledge.com/Transformational-Leadership/Bass-Riggio/p/book/9780805847628 (accessed on 5 July 2022).
55. Weisinger, H., (1998). *Emotional Intelligence at Work: The Untapped Edge for Success.* San Francisco: Jossey-Brass.
56. Goleman, D., Boyatzis, R. E., & McKee, A., (2002). *Primal Leadership: Realizing the Power of Emotional Intelligence.* Boston: Harvard Business School Press.
57. Hong, Y., Catano, V. M., & Liao, H., (2011). Leader emergence: The role of emotional intelligence and motivation to lead. *Leadership & Organization Development Journal, 32*(4), 320–343.
58. Hur, Y., Van, D. B. P. T., & Wilderom, C. P., (2011). Transformational leadership as a mediator between emotional intelligence and team outcomes. *The Leadership Quarterly, 22*(4), 591–603.
59. Bass, B. M., (1985). *Leadership and Performance Beyond Expectations.* New York: The Free Press.
60. Burns, J. M., (1978). *Leadership.* New York: Harper and Row.
61. Ross, S. M., & Offermann, L. R., (1997). Transformational leaders: Measurement of personality attributes and workgroup performance. *Personality and Social Psychology Bulletin, 23*(10), 1078–1086.
62. Goleman, D., (1995). *Emotional Intelligence.* New York: Bantam Books.
63. Mandell, B., & Pherwani, S., (2003). Relationship between emotional intelligence and transformational leadership style: A gender comparison. *Journal of Business and Psychology, 17*(3), 387–404.
64. Kerr, R., Garvin, J., Heaton, N., & Boyle, E., (2006). Emotional intelligence and leadership effectiveness. *Leadership & Organization Development Journal, 27*(4), 265–279.
65. Madera, J. M., & Smith, D. B., (2009). The effects of leader negative emotions on evaluations of leadership in a crisis situation: The role of anger and sadness. *The Leadership Quarterly, 20*(2), 103–114.
66. Thiel, C. E., Connelly, S., & Griffith, J. A., (2012). Leadership and emotion management for complex tasks: Different emotions, different strategies. *The Leadership Quarterly, 23*(3), 517–533.
67. Newcombe, M. J., & Ashkanasy, N. M., (2002). The role of affect and affective congruence in perceptions of leaders: An experimental study. *The Leadership Quarterly, 13*(5), 601–614.

34   *Emotional Intelligence for Leadership Effectiveness*

68. Weiss, H. M., & Cropanzano, R., (1996). Affective events theory: A theoretical discussion of the structure, causes and consequences experiences at work. In Staw, B. M., & Cummings, L. L., (eds.), *Research in Organizational Behavior* (pp. 1–74). Greenwich, CT: JAI Press.
69. De Vries, M. F. K., (2006). *The Leader on the Couch: A Clinical Approach to Changing People and Organizations.* John Wiley & Sons.
70. Pirola-Merlo, A., Härtel, C., Mann, L., & Hirst, G., (2002). How leaders influence the impact of affective events on team climate and performance in R&D teams. *The Leadership Quarterly, 13*(5), 561–581.
71. McColl-Kennedy, J. R., & Anderson, R. D., (2002). Impact of leadership style and emotions on subordinate performance. *The Leadership Quarterly, 13*(5), 545–559.
72. Torrence, B. S., & Connelly, S., (2019). Emotion regulation tendencies and leadership performance: An examination of cognitive and behavioral regulation strategies. *Frontiers in Psychology, 10*, 1486.
73. Bono, J. E., & Ilies, R., (2006). Charisma, positive emotions and mood contagion. *The Leadership Quarterly, 17*(4), 317–334.
74. Hatfield, E., Cacioppo, J. T., & Rapson, R. L., (1994). *Emotional Contagion.* Cambridge: Cambridge University Press.
75. Van, K. G. A., Homan, A. C., Beersma, B., Van, K. D., Van, K. B., & Damen, F., (2009). Searing sentiment or cold calculation? The effects of leader emotional displays on team performance depend on follower epistemic motivation. *Academy of Management Journal, 52*(3), 562–580.
76. Bass, B. M., & Stogdill, R. M., (1990). *Bass & Stogdill's Handbook of Leadership: Theory, Research, and Managerial applications.* New York: Free Press.
77. Shiao, S. P. K., Hutto, N. N., Andrews, C. M., & Lucke, K. T., (2011). Creating a healthy work environment for nursing practice and education: Leadership impact on nursing and healthcare work environment. *International Journal of Work Organization and Emotion, 4*(3, 4), 301–321.
78. George, J. M., (2000). Emotions and leadership: The role of emotional intelligence. *Human Relations, 53*(8), 1027–1055.
79. Norris, F. H., Friedman, M. J., Watson, P. J., Byrne, C. M., Diaz, E., & Kaniasty, K., (2002). 60,000 disaster victims speak: Part I. An empirical review of the empirical literature, 1981–2001. *Psychiatry: Interpersonal and Biological Processes, 65*(3), 207–239.
80. Trevor, C. O., & Nyberg, A. J., (2008). Keeping your headcount when all about you are losing theirs: Downsizing, voluntary turnover rates, and the moderating role of HR practices. *Academy of Management Journal, 51*(2), 259–276.
81. Fitzpatrick, K. M., Harris, C., & Drawve, G., (2020). Fear of COVID-19 and the mental health consequences in America. *Psychological Trauma: Theory, Research, Practice, and Policy, 12*(S1), S17–S21.
82. Kramer, A., & Kramer, K. Z., (2020). The potential impact of the COVID-19 pandemic on occupational status, work from home, and occupational mobility. *Journal of Vocational Behavior, 119*, 103442.
83. Kniffin, K. M., Narayanan, J., Anseel, F., Antonakis, J., Ashford, S. P., Bakker, A. B., & Vugt, M. V., (2021). COVID-19 and the workplace: Implications, issues, and

insights for future research and action. *American Psychologist, 76*(1), 63–77. https://doi.org/10.1037/amp0000716.

84. Gupta, S., (2021). *6 Ways to Support Employee Well-Being Amid COVID-19 Era.* India Today. Retrieved from: https://www.indiatoday.in/education-today/featurephilia/story/6-ways-to-support-employee-well-being-amid-covid-19-era-1805668-2021-05-22 (accessed on 5 July 2022).

85. Rodriguez, S., (2020). *Facebook Is Offering Employees up to a Month of Paid Leave to Care for Sick Family Members.* CNBC. Retrieved from: https://www.cnbc.com/2020/03/23/facebook-staff-get-30-days-paid-leave-to-care-for-coronavirus-sick.html (accessed on 5 July 2022).

86. Phadnis, S., (2021). *Companies take COVID Relief Initiatives for Employees.* The Times of India. Retrieved from: https://timesofindia.indiatimes.com/business/india-business/companies-take-covid-relief-initiatives-for-employees/articleshow/82513889.cms (accessed on 5 July 2022).

87. Mallet, V., & Dombey, D., (2020). *France, Spain and UK Unleash Rescue Packages to Help Companies.* Financial Times. Retrieved from: https://www.ft.com/content/7eb398ac-6839-11ea-800d-da70cff6e4d3 (accessed on 5 July 2022).

88. Yao, Y. H., Fan, Y. Y., Guo, Y. X., & Li, Y., (2014). Leadership, work stress and employee behavior. *Chinese Management Studies, 8*(1), 109–126.

CHAPTER 3

The Impact of Emotional Intelligence and Leadership in Pandemic Times

MARIANA REZENDE ALVES DE OLIVEIRA,[1]
MUBASHIR MAJID BABA,[2] and JOSÉ APARECIDO DA SILVA[3]

[1]*Master's Student in Psychobiology–FFCLRP, USP–University of São Paulo at Ribeirão Preto, Virtual Laboratory of Affective, Cognitive, and Behavioral Neuropsychometry–LAVINACC,*
E-mail: mari-rao@hotmail.com

[2]*Department of Management Studies, North Campus, University of Kashmir, Srinagar, Jammu and Kashmir, India; Former Project Fellow, IIM Lucknow, Uttar Pradesh, India*

[3]*Full Professor of 4P: Psychometrics, Psychophysics, Perception, and Pain Department of Psychology, University of São Paulo at Ribeirão Preto, Brazil*

ABSTRACT

Sudden change of routine, high risk of infection, fear of illness, social isolation, and implementation of new behaviors and attitudes to the population in general have been factors that contributed significantly to the increase of panic attacks, anxiety, depression, stress, and fear. Faced with this scenario of profound mental illness of the population, the recommended response of intervention in psychological crisis during COVID-19, the focus was on maintaining emotional stability and coping methods. At the national level, the same mental illness has been observed

---

Emotional Intelligence for Leadership Effectiveness: Management Opportunities and Challenges During Times of Crisis. Mubashir Majid Baba, Chitra Krishnan, & Fatma Nasser Al-Harthy (Eds.)
© 2023 Apple Academic Press, Inc. Co-published with CRC Press (Taylor & Francis)

in the Brazilian population. However, preventive measures for mental health care, promotion of information and coping strategies on stress and emotional intelligence (EI) were not disclosed. Some studies relate EI to the reduction of exhaustion, stress, and better performance at work, especially among health professionals. Thus, this chapter analyzed the EI other affective factors (such as happiness, psychological stress, resilience, and non-somatic pain) and other affective factors six months into the COVID-19 pandemic.

3.1 INTRODUCTION

As a result of this epidemic, the mental health and physical health of the world's population has been severely affected. Sudden changes in daily routine, high risk of infection, fear of illness, social isolation, and implementation of new behaviors and attitudes to the general population have been factors that have contributed significantly to the increase of panic attacks, anxiety, depression, stress, and fear [1]. Faced with this scenario of profound mental illness of the population, the recommended response of intervention in psychological crisis for the COVID-19 epidemic was focused on maintaining emotional stability, fighting dread, measuring pain, and improving coping techniques [2].

During the period of social isolation, the world population had to readjust and develop a certain resilience to face this moment. Several people found themselves outside their support network and isolated from their friends and family. Thus, people around the world, imposed to social isolation in favor of individual and collective health, faced insomnia, symptoms of anxiety, suicidal ideas, and emotional dysregulation for the first time in their lives [3]. The emotional responses of the population during a pandemic can include fear, uncertainty about the future, negative social attitudes, and behaviors due to a wrong view of the pandemic scenario and a distorted perception of risk of contamination [4].

People's responses to COVID-19, since the beginning of the pandemic, have been shaped according to the opinion of their respective leaders and are a reflection of their political preferences [3]. According to the authors [3], if the authorities have the intelligence to connect empathetically with the population, treat them with respect and educate them about all the measures that should be taken in the moment of crisis, instead of simply forcing a lockdown, it is more likely that the population will identify a

The Impact of Emotional Intelligence and Leadership 39

sense of collectivity and will adhere to the measures without so much suffering.

World leaders had to act in order to control panic and calm the population with security measures and evidence that showed that they were working to try to control the health and economy of each country. Such attitudes demand emotional intelligence (EI) and attitudes that leaders would need to have to make an effective control of the pandemic, they are to represent the population; take all measures by the population; elaborate and incorporate a sense of collectivity [3, 5]. Some studies have shown that this capacity may be linked to better control of the pandemic and the population's adherence to security measures [3, 5, 6].

3.2 LITERATURE APPROACH

During COVID-19 EI plays a crucial part in properly coping with daily environmental stresses: emotional control, relationships, self-awareness, and effective communication [1]. The concept was formally defined by Salovey and Mayer in 1990 as "the ability to monitor one's own and others' feelings and emotions, to discriminate among them and to use this information to guide one's thinking and actions" [7]. Later, in 1998, Goleman [8] defined EI as "the capacity for recognizing our own feelings and those of others, for motivating ourselves and for managing emotions well in ourselves and in our relationships" [8].

"It is a dynamic condition that can be learned by a leader or an organization" [9]. EI consists of two general areas: Personal Competence (self-management and self-awareness) and Social Competence (interpersonal relationships [5]. Goleman [10] and Bar-On [11] consider self-awareness the most significant feature of EI [1, 10, 11]. As a result, leaders must be able to understand the emotional reality and cultural norms that drive the conduct of their organizations [9].

3.3 METHODS AND INSTRUMENTS

A cross-sectional survey was done from August 20th to October 23rd, 2020, to assess EI and other psychological factors, such as happiness, psychological stress, resilience, and non-somatic pain, among other things. Google Forms was used to collect the data, which was then forwarded to participants via social media and e-mail. This research is part of the project titled Physical, Psychological, and Cognitive Reactions to COVID-19 (11th module).

In addition, participants were needed to sign a Participant Consent Form that outlined the study's objectives and nature, as well as a statement that they could withdraw at any moment. In total, 110 Brazilians participated in the poll, which was also the final sample size for the study's analysis.

3.4 INSTRUMENTS

EI was assessed using the Emotional Intelligence Scale (EIS) *during COVID-19 Pandemic* [1]. 5-point Likert scale was used from 1 to 5 (5, = strongly agree, 4 = agree, 3 = neutral, 2 = disagree and 1 = strongly disagree) [1]. It is a scale with a high reliability (Cronbach's alpha = 0.768) [1]. To better fit our study, we used the items 1 to 22 (the full scale has 28 items). The author gave us permission to use the weighing scale. The items were translated and converted to the Brazilian Portuguese language by our team of translators.

We assessed resilience using the brief resilience scale (BRS) [12]. It consists in a 5-point Likert-type 6 items scale where 1 = *strongly disagree*, 2 = *disagree*, 3 = *neutral*, 4 = *agree* and 5 = *strongly agree* [12]. The score ranges from 6 to 30. It demonstrates good internal consistency (Cronbach's alpha = 0.80).

We measured happiness during the pandemic with the *Oxford Happiness Questionnaire* (OHQ) [13]. It consists of a 29-item questionnaire presented in a Likert format in six categories of responses (*1 = strongly disagree, 2 = moderately disagree, 3 = slightly disagree, 4 = slightly agree, 5 = moderately agree, 6 = strongly agree*) [13]. The questionnaire has a high and strong reliability and internal consistency (Cronbach's alpha = 0.91).

We also used the 6-item *Kessler Psychological Distress Scale* (K6) [14] to measure non-specific psychological distress. It is a five-point scale (0 = none of the time, 1 = a little of the time, 2 = some of the time, 3 = most of the time, 4 = all of the time) that the responses range from 0–24 [14]. It has confirmed excellent internal consistency and reliability (Cronbach's alpha = 0.89).

To quantify emotional and subjective pain we used *Non-Somatic Pain Scale*, developed by one of the authors of this study, José Aparecido. It

The Impact of Emotional Intelligence and Leadership 41

consists in a three-item scale presented in a Likert format in four categories of responses (1 = never, 2 = sometimes, 3 = most of the time, 4 = All of the time). It has demonstrated great internal consistency and reliability (Cronbach's alpha = 0.87).

Afterwards, the participants completed a socio-demographic survey that included questions on their professions, social isolation, and work status as well as their emotional state during quarantine.

3.5 STATISTICAL ANALYSIS

All data were analyzed using the descriptive statistics method in IBM SPSS Statistics 23.

3.6 RESULTS AND DISCUSSION

To characterize the sample, we assessed some factors (age, gender, job, education level, social isolation, perceived happiness, and non-physical pain) in the socio-demographic survey, as shown in Table 3.1.

In our sample, 89 participants (80.91%) were female, with a mean age of 36, 9 years old, with a complete college degree (40%), low perceived happiness (51.82%) and non-physical pain sensation (55.45%) during the COVID-19 pandemic. This profile may be significant because, as the literature has been showing since the beginning of the pandemic, women are showing higher levels of stress and psychological distress as they try to conciliate their careers, house tasks, and personal life during social isolation times [15, 16].

Despite the lack of an official lockdown in Brazil, most of our sample kept self-isolation (55.45%). It is necessary to consider the fact that by the time this research took place, most cities and states had started to reopen non-essential services, and by September, some schools had reopened [16, 17].

Scientific understandings of the pandemic scenario of COVID-19 undergo daily changes, causing fear, uncertainty, and even mistrust in the population [18]. In order to contain the damage of this pandemic, WHO promoted preventive behavioral measures, which were imposed on the world's population from day to day. The night: social distance; hygiene practices; wearing masks; among others [19].

42   *Emotional Intelligence for Leadership Effectiveness*

TABLE 3.1   Characteristics of the Participants (n = 110)

Variable	N	Percentage (%)
Female	89	80.91%
Male	21	19.09%
Age (Years)		
Mean	36.9	–
Median	35	–
Mode	27	–
Job		
Total People*	**60**	54.55%
Professor/teacher	34	30.91%
Health professional	24	21.82%
Health professional (frontline)	2	1.82%
Education Level		
Secondary school (complete)	2	1.82%
High school (complete)	16	14.55%
College degree (complete)	44	40%
College degree (incomplete)	6	5.45%
Postgraduate studies (complete)	39	35.45%
Postgraduate studies (incomplete)	3	2.73%
Non-Physical Pain		
Yes	61	55.45%
No	49	44.55%
Social Isolation (6 Months into the Pandemic)		
Yes	85	77.27%
No	25	22.73%
Perceived Happiness During the Pandemic		
Happy	26	23.64%
A little happy	57	51.82%
Nothing has changed	20	18.2%
Very unhappy	7	6.34%

Unlike what has been shown in the international literature for other countries [16, 20, 21], Brazilian leaders did not adopt official measures of lockdown or effective prevention policies. According to literature [1, 2, 6], the lack of actions that would cause a sense of collectivity in the population, can be considered as a was a low capacity for EI for pandemic control.

Preventive measures for mental health care, promotion of information and coping strategies on stress and EI were not disclosed in Brazil. In Table 3.2, it is possible to see the gender frequency scores of the scales that were used. In two (K6 and Non-Somatic Pain Scale) of the four scales used, women's mean scores were significantly higher than those of men. In Kessler Psychological Distress Scale (K6) 38.20% of women sample are said to have a severe level of distress, while 52.38% of men sample are categorized on the normal level. Also, the female sample presents a higher mean and maximum scores (6.51 and 12) than the male sample (5.3 and 3) in the Non-Somatic Pain Scale (which contains affirmatives about deep sadness and agony).

These results indicate that the female sample is more likely to be associated with increased distress. It is in line with others results in the literature [16, 22–24]. Some studies relate EI to the reduction of exhaustion, stress, and better performance at work, especially in health professionals [25, 26]. Women are being more affected by the COVID-19 pandemic than men. Meanwhile, on the BRS the female sample (3.23) presented a mean score lower than the male (3.58) and total sample (3.30). This result may be considered as a factor to why women are being the most affected. Considering the excessive load of work at their jobs, at home with their family, the stress and anxiety symptoms appear more frequently [27], since the intensity of affective reactions correlates with internalizing and externalizing symptomatology of anxiety and distress [26].

On the EI scale [1] and the OHQ [12] it is still possible to point the gender difference. The male sample presents a higher mean score (87.19) than the female sample (82.51) in the EI scale. This result, despite both samples reached a high score of 110, may explain the differences between the scores obtained on the K6 scale. As EI is considered to make a person more capable of dealing with adversities and their own and other's emotions, it was expected that higher scores in EI would relate to higher levels of happiness and lower levels of distress.

EI is adversely connected with anxiety, depression, and stress, according to several research [28, 29]. Also, studies have shown that EI can assist people notice and reduce the severity of unpleasant emotions [30].

This being said, it is possible to relate the higher performance of the male sample in the EI Scale with lower scores on the K6, as well the significantly lower performance on EI scale and higher scores on K6 from the female sample. In this study we found a strong, negative correlation between the EI Scale During COVID-19 Pandemic and the 6-item Kessler.

44   *Emotional Intelligence for Leadership Effectiveness*

TABLE 3.2 Gender Frequency and Statistical Scores for OHQ, K6, EI Scale, BRS, and Non-Somatic Pain Scale

	Women	Men	Sample
OHQ Statistics			
Mean	3.95	4.26	4.04
Median	4.06	4.41	4.18
Mode	4.03	5.41	4.37
Min	2.34	2.65	2.34
Max	5.59	5.41	5.59
OHQ Frequency			
Not happy (1–2)	10 (10.11%)	3 (14.29%)	12 (10.91%)
Not particularly happy or unhappy (3–4)	76 (85.39%)	15 (71.42%)	91 (82.73%)
Very happy (5–6)	4 (4.50%)	3 (14.29%)	7 (6.36%)
K6 Frequency			
Normal (0–7)	31 (34.83%)	11 (52.38%)	42 (38.18%)
Mild (8–12)	24 (26.96%)	5 (23.81%)	12 (26.36%)
Severe (13–24)	34 (38.20%)	5 (23.81%)	39 (35.45%)
EI Scale Statistics			
Mean	82.51	87.19	83.4
Median	82	84	82.5
Mode	82	76	75
Min	53	65	53
Max	110	110	110
BRS Statistics			
Mean	3.23	3.58	3.30
Median	3.33	3.5	3.33
Mode	4	5	4
Min	1	1.83	1
Max	5	5	5
Non-Somatic Pain			
Mean	6.51	5.3	6.29
Median	6	5	6
Mode	6	3	6
Min	3	3	3
Max	12	3	12

Psychological Distress Scale (K6) (r = 0.639; p < 0.001), as it is shown in Figure 3.1. The negative correlations mean that in the presence of a low EI, a higher level of distress may appear, which is aligned with the literature about EI and negative emotions [29, 30].

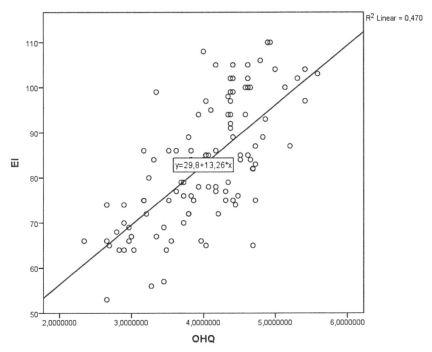

FIGURE 3.1 Emotional intelligence scale during COVID-19 pandemic (y) and the 6-item Kessler psychological distress scale (K6) (x) and Pearson coefficient using scatter diagram.

The results in Figure 3.2 also showed the EI scale during COVID-19 pandemic and the Oxford happiness questionnaire (OHQ) demonstrated a substantial, positive association. (r = 0.686; p < 0.001). This indicates that higher EI may lead to higher levels of happiness.

Table 3.3 shows the statistics scores for the EI scale during COVID-19 Pandemic and the 6-item Kessler psychological distress scale (K6) for health professional sample (24) and professors/teachers' sample (33). Health professionals and teachers presented similar results and mean scores on both scales. The high scores on EI Scale justify the lower scores on the K6 scale.

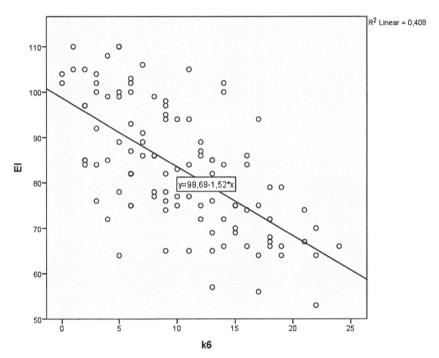

FIGURE 3.2 Emotional intelligence scale during COVID-19 pandemic (y) and the Oxford happiness questionnaire (x) and Pearson coefficient using scatter diagram.

TABLE 3.3 Statistics Scores for the Emotional Intelligence Scale During COVID-19 Pandemic and the 6-Item Kessler Psychological Distress Scale

	Health Professional	Professor/Teacher	Sample
	K6 Frequency		
Normal (0–7)	12 (50%)	19 (57.58%)	42 (38.18%)
Mild (8–12)	6 (25%)	7 (21.21%)	12 (26.36%)
Severe (13–24)	6 (25%)	7 (21.21%)	39 (35.45%)
	EI Scale Statistics		
Mean	88.54	89.18	83.4
Median	87.5	87	82.5
Mode	82	82	75
Min	70	57	53
Max	108	110	110

The Impact of Emotional Intelligence and Leadership

Several studies have shown that health care workers are most prone to suffer from psychological symptoms (such as panic syndrome, anxiety, post-traumatic stress disorder and depression) [1, 16, 31]. Overwork, frustration, discrimination, isolation, patients with bad emotions, lack of touch with their families and weariness have been among the challenges faced by medical personnel in Wuhan during the COVID-19 pandemic [31].

Some studies also showed that schoolteachers and university professors have been presenting significant levels of stress, anxiety, and burnout since the beginning of online classes [32–34]. It is important to outline that by the time of this survey, schools were still closed or had just reopened. This way, we can infer that the professors/teacher's sample had been for 5 months teaching remotely and were significant adapted to the new scenario.

Half of health professional sample presented mild and severe distress (25% and 25%, respectively), which may correlate with the pandemic situation at the time. By the time of this study, As of September 12th, 2020, Brazil had 20,126 confirmed cases per 1 million of the population and 613 by COVID-19 [35, 36].

Leading with EI would establish a common goal and an acknowledgement of the factors over which cannot be controlled [6]. The fact that leaders do not make consensual decisions about the pandemic and what security measures should be implemented, ends up leaving health professionals and teachers working in a dangerous and infectious environment, which may explain some levels of psychological distress.

3.7 CONCLUSIONS

A few drawbacks of this study should be noted. Although we reached participants in 13 different U.S. states, this poll should not be viewed as representative of the entire country's population. In this study, a limited sample was examined to determine the impact of the pandemic. The results can be related to the way the Brazilian leaders has been dealing with the COVID-19 pandemic, which correlates with the literature about EI and leadership. A number of studies have shown that diseases are followed by significant individual and social psychosocial effects, which can be far more damaging than the disease itself [37]. World population has been adversely affected by the COVID-19 epidemic, which has spread throughout the world. As a result, the pandemic had a greater impact on countries whose leaders lacked EI when dealing with the problem [3, 16].

A number of studies have shown that EI is necessary for dealing with COVID-19-induced stress, anxiety, and other unwanted effects [1, 3, 30]. In Brazil, a significant decrease in the populations mental health has been observed, mainly in people who have lost a relative to COVID-19 or had some high economic loss, also affected by gender. However, there has been a wave of negationism during the pandemic and a complete lack of EI by the government, which is still negatively affecting the population.

KEYWORDS

- **Brazil**
- **brief resilience scale**
- **coronavirus disease 2019**
- **emotional intelligence**
- **emotional intelligence inventory**
- **Kessler psychological distress scale**
- **Oxford happiness questionnaire**

REFERENCES

1. Baba, M. M., (2020). Navigating COVID-19 with emotional intelligence. *International Journal of Social Psychiatry, 66*(8), 810–820. https://doi.org/10.1177/0020764020934519.
2. Morón, M., & Biolik-Moroń, M., (2021). Trait emotional intelligence and emotional experiences during the COVID-19 pandemic outbreak in Poland: A daily diary study. *Personal. Individ. Differ., 168*, 110348. http://doi.org/10.1016/j.paid.2020.110348
3. Jetten, J., Reicher, S. D., & Cruwys, T., (2020). *Together Apart: The Psychology of COVID-19* (1st edn.). SAGE Publications.
4. Taylor, S., (2019). *The Psychology of Pandemics: Preparing for the Next Global Outbreak of Infectious Disease.* Cambridge Scholars Publishing.
5. Goleman, D., Boyatzis, R., & McKee, A., (2004). *Primal Leadership: Learning to Lead with Emotional Intelligence.* Boston, MA: Harvard Business School Press.
6. Ward, H. B., (2020). Resident leadership in the era of COVID-19: Harnessing emotional intelligence. *Academic Medicine: Journal of the Association of American Medical Colleges, 95*(10), 1521–1523. https://doi.org/10.1097/ACM.0000000000003558

The Impact of Emotional Intelligence and Leadership

49

7. Salovey, P., & Mayer, J. D., (1990). Emotional intelligence. *Imagination, Cognition and Personality, 9*(3), 185–211. https://doi.org/10.2190/DUGG-P24E-52WK-6CDG.

8. Goleman, D., (1998). *Working with Emotional Intelligence.* Bantam Books.

9. Goleman, D., Boyatzis, R. E., & McKee, A., (2002). *Primal Leadership: Realizing the Power of Emotional Intelligence.* Bantam Books. New York, NY.

10. Goleman, D., (1996). *Emotional Intelligence: Why It Can Matter More Than IQ.* Bloomsbury Publishing.

11. Bar-On, R., (1997). *The Emotional Intelligence Inventory (EQ-I): Technical Manual.* Multi-Health Systems.

12. Krishnan, C., Goel, R., Singh, G., Bajpai, C., & Malik, P., (2017). Emotional intelligence: A study on academic professionals. *Pertanika Journal of Social Sciences & Humanities.*

13. Smith, B. W., Dalen, J., Wiggins, K., et al., (2008). The brief resilience scale: Assessing the ability to bounce back. *Int. J. Behav. Med., 15,* 194–200. https://doi.org/10.1080/10705500802222972.

14. Hills, P., & Argyle, M., (2002). The oxford happiness questionnaire: A compact scale for the measurement of psychological well-being. *Personality and Individual Differences, 33,* 1073–1082. https://doi.org/10.1016/S0191-8869(01)00213-6.

15. Kessler, R., andrews, G., Colpe, L., Hiripi, E., Mroczek, D., Normand, S., .Zaslavsky, A., (2002). Short screening scales to monitor population prevalences and trends in non-specific psychological distress. *Psychological Medicine, 32*(6), 959–976. doi: 10.1017/S0033291702006074.

16. Zysberg, L., & Zisberg, A., (2020). Days of worry: Emotional intelligence and social support mediate worry in the COVID-19 pandemic. *Journal of Health Psychology.* https://doi.org/10.1177/1359105320949935.

17. Passos, L., Prazeres, F., Teixeira, A., & Martins, C., (2020). Impact on mental health due to COVID-19 pandemic: Cross-sectional study in Portugal and Brazil. *International Journal of Environmental Research and Public Health, 17*(18), 6794. https://doi.org/10.3390/ijerph17186794.

18. Government of the State of Sao Paulo. https://www.saopaulo.sp.gov.br/sala-de-imprensa/release/governo-de-sp-anuncia-retomada-das-aulas-para-8-de-setembro/ (accessed on 5 July 2022).

19. Vasconcellos-Silva, P. R., & Castiel, L. D., (2020). COVID-19, fake news and the sleep of communicative reason generating monsters: The narrative of risks and the risks of narratives. *Public Health Notebooks, 36*(7), e00101920. https://doi.org/10.1590/0102-311x00101920.

20. World Health Organization (2020). *Coronavirus Disease (COVID-19): Situation Report 106.* World Health Organization. Disponível em: https://www.who.int/emergencies/diseases/novel-coronavirus-2019/situation-reports (accessed on 5 July 2022).

21. Brauner, J. M., Mindermann, S., Sharma, M., Johnston, D., Salvatier, J., Gaven˘ciak, T., et al., (2020). Inferring the effectiveness of government interventions against COVID-19. *Science.* doi: 10.1126/science.abd9338.

22. Majumdar, P., Biswas, A., & Sahu, S., (2020). COVID-19 pandemic and lockdown: Cause of sleep disruption, depression, somatic pain, and increased screen exposure of office workers and students of India. *Chronobiology International, 37*(8), 1191–1200. https://www.tandfonline.com/doi/abs/10.1080/07420528.2020.1786107?journalCode=icbi20.

23. Mazza, C., Ricci, E., Biondi, S., Colasanti, M., Ferracuti, S., Napoli, C., & Roma, P., (2020). A nationwide survey of psychological distress among Italian people during the COVID-19 pandemic: Immediate psychological responses and associated factors. *International Journal of Environmental Research and Public Health, 17*, 3165. https://doi.org/10.3390/ijerph17093165.

24. Gómez-Salgado, J., Andrés-Villas, M., Domínguez-Salas, S., Díaz-Milanés, D., & Ruiz-Frutos, C., (2020). Related health factors of psychological distress during the COVID-19 pandemic in Spain. *International Journal of Environmental Research and Public Health, 17*(11), 3947. https://doi.org/10.3390/ijerph17113947.

25. Choi, E., Hui, B., & Wan, E., (2020). Depression and anxiety in Hong Kong during COVID-19. *International Journal of Environmental Research and Public Health, 17*(10), 3740. https://doi.org/10.3390/ijerph17103740.

26. Soto-Rubio, A., Giménez-Espert, M., & Prado-Gascó, V., (2020). Effect of emotional intelligence and psychosocial risks on burnout, job satisfaction, and nurses' health during the COVID-19 pandemic. *International Journal of Environmental Research and Public Health, 17*(21), 7998. https://doi.org/10.3390/ijerph17217998.

27. Almeida, M., Shrestha, A. D., Stojanac, D., & Miller, L. J., (2020). The impact of the COVID-19 pandemic on women's mental health. *Archives of Women's Mental Health, 23*(6), 741–748. https://doi.org/10.1007/s00737-020-01092-2.

28. Wenham, C., Smith, J., Davies, S. E., Feng, H., Grépin, K. A., Harman, S., Herten-Crabb, A., & Morgan, R., (2020). Women are most affected by pandemics-lessons from past outbreaks. *Nature, 583*(7815), 194–198. https://doi.org/10.1038/d41586-020-02006-z.

29. Cejudo, J., Rodrigo-Ruiz, D., López-Delgado, M. L., & Losada, L., (2018). Emotional intelligence and its relationship with levels of social anxiety and stress in adolescents. *International Journal of Environmental Research and Public Health, 15*(6), 1073. https://doi.org/10.3390/ijerph15061073.

30. Sun, H., Wang, S., Wang, W., Han, G., Liu, Z., Wu, Q., & Pang, X., (2021). Correlation between emotional intelligence and negative emotions of front-line nurses during the COVID-19 epidemic: A cross-sectional study. *J. Clin. Nurs., 30*, 385–396. https://doi.org/10.1111/jocn.15548.

31. Molero, J. M., Pérez-Fuentes, M., Oropesa, R. N. F., Simón, M. M., & Gázquez, L. J. J., (2019). Self-efficacy and emotional intelligence as predictors of perceived stress in nursing professionals. *Medicine (Kaunas, Lithuania), 55*(6), 237. https://doi.org/10.3390/medicina55060237.

32. Torales, J., O'Higgins, M., Castaldelli-Maia, J. M., & Ventriglio, A., (2020). The outbreak of COVID-19 coronavirus and its impact on global mental health. *The International Journal of Social Psychiatry, 66*(4), 317–320. https://doi.org/10.1177/0020764020915212.

33. Akour, A., Al-Tammemi, A. B., Barakat, M., Kanj, R., Fakhouri, H. N., Malkawi, A., & Musleh, G., (2020). The impact of the COVID-19 pandemic and emergency distance teaching on the psychological status of university teachers: A cross-sectional study in Jordan. *The American Journal of Tropical Medicine and Hygiene, 103*(6), 2391–2399. https://doi.org/10.4269/ajtmh.20-0877.

34. Kim, L. E., & Asbury, K., (2020). 'Like a rug had been pulled from under you': The impact of COVID-19 on teachers in England during the first six weeks of the UK

lockdown. *The British Journal of Educational Psychology, 90*(4), 1062–1083. https://doi.org/10.1111/bjep.12381.

35. UNESCO, (2020b). *Adverse Consequences of School Closures*. UNESCO. Retrieved from: https://en.unesco.org/covid19/educationresponse/consequences (accessed on 5 July 2022).

36. Worldometer, (2020). *Coronavirus Cas*es. Available online: https://www.worldometers.info/coronavirus/? (accessed on 5 July 2022).

37. Ministry of Health, (2020). *Panel of Cases of Coronavirus Disease 2019 (COVID-19) in Brazil by the Ministry of Health*. URL. https://covid.saude.gov.br/ (accessed on 5 July 2022).

38. Ornell, F., Halpern, S. C., Kessler, F. H. P., & Narvaez, J. C. M., (2020). The impact of the COVID-19 pandemic on the mental health of healthcare professionals. *Cad Saude Publica, 36*(4), e00063520. doi: 10.1590/0102-311X00063520. PMID: 32374807.

CHAPTER 4

Emotional Intelligence and Its Management in the Conflicting Factors in the Pandemic by Using a Mediative Fuzzy Logic System

NITESH DHIMAN,[1] MEETIKA SHARMA,[2] and M. K. SHARMA[1]

[1]Department of Mathematics, Chaudhary Charan Singh University, Meerut–250004, Uttar Pradesh, India, E-mail: drmukeshsharma@gmail.com (M. K. Sharma)

[2]Department of English, J.L.M. College, OFC, Muradnagar, Ghaziabad–201206, Uttar Pradesh, India

ABSTRACT

This chapter explores that how mediative fuzzy logic be used in emotional intelligence (EI) and its management of the conflicting factors in the pandemic. Mediative fuzzy logic is applicable as an extension of intuitionistic fuzzy logic to control the conflicting factors in pandemic period. It can manage the contradictory information, which is very useful to manage the emotional quotient as compared to intelligent quotient. The present work is often based on measurements of the Intelligence Quotient (IQ) levels such as mathematical computation. The factors to the confliction at home during pandemic are categories into five major categories as financial crisis, psychological behavior, interpersonal effects, clinical implication for health issues, and education loss. In addition, the relationship between the input factors and the conflicting factors has also

---

Emotional Intelligence for Leadership Effectiveness: Management Opportunities and Challenges During Times of Crisis. Mubashir Majid Baba, Chitra Krishnan, & Fatma Nasser Al-Harthy (Eds.)
© 2023 Apple Academic Press, Inc. Co-published with CRC Press (Taylor & Francis)

54   *Emotional Intelligence for Leadership Effectiveness*

been defined to infer the output of the mediative based fuzzy inference system.

4.1 INTRODUCTION

Pandemic multiplies many existing conflicts. One of them is knowledge of the world around. Human beings acquire knowledge through their experiences, communication, society, family, and education. Knowledge plays a central role in managing the assessment and relevance of the conflicting factors. But knowledge is affected by the emotional intelligence (EI). The problem with the emotional knowledge is that its most of the part is perception based. Perceptions and especially perception of the possibilities and the probabilities of the factors are intrinsically imprecise and reflect the fact that human sensory organs, and ultimately the brain, have bounded. They lose the ability to resolve the conflicting factors which may arise during their stay-at-home stay in the pandemic situations. Pandemics are basically the global epidemic that occurs over a very wide area, crossing the natural boundary and usually affecting a large amount of society. Pandemics are having a classical, social, and economic impact on the lives of human beings. The pandemic changes the lives of people. The economic and social disruption due to the pandemic is divesting the people and begets conflict among the people during their stay at home.

There are various factors which gives birth to the conflict at home during pandemic like financial crisis (F), psychological behavior (P), interpersonal effects (I), clinical implication for health issues (C), and education loss (E).

Imprecision of perception stands in before the eyes of conventional technology in the measurement of conflicting factors. These techniques are based on the bivalent logic and probability theory. Another complication is: the too much knowledge about the pandemic. The human brain receives many kinds of information from various sources reliable and unreliable as well. This is one of the biggest obstacles and it centers upon the concept of relevance. As perception varies from man to man, so the conflicts may vary.

The third obstacle is the emotional quotient (EQ). The EI is the potential to figure out, use, and maintain its own emotion in a positive sense to cure stress, self-awareness, and to communicate effectively. It may have empathy for others to overcome the challenges. The conflict during stay at home in the pandemic situation may occur. There are mainly five attributes which define the EI, i.e., self-management, self-awareness, handling relationship, self-motivation, and empathy. There are different ways to increase the level of EI. Various techniques are also available to measure the EQ at different levels, i.e., perceiving emotions, reasoning with emotions, understanding with emotions, and managing emotions.

In this present proposal, we will formulate a mediative fuzzy logic model in the management of pandemic days over these three obstacles, i.e., knowledge perception, conflicting factors, and EI.

There are two different ways for the assessment of the model:

i. Semantic; and
ii. Statistical.

Research was conducted (Sir Sanford Fleming College, Canada) in 2009. This study shows that the major reason of student withdrawal from their respective colleges was related to two factors, i.e., social, and emotional. To discriminates between the emotions level and to use emotional knowledge to handle the thinking and behavior; a study [2] was conducted in 2009. Austin's [3] gave a study based on a limited correlation between academic scores and EI scores among students in the medical field. Many other studies [4–22] on EI were also conducted so far.

In real-life situations, when human intervention is being quantized then the statistical methods have no relevance. To deal with such a situation, a logic-based method is required. Fuzzy logic [23] is a logic that deals with the real-life situation and works like human intelligence. Fuzzy logic is a rule-based inference model that can rely on the empirical happening of an entrepreneur, especially usable to hold escapades promoter accomplishment. Fuzzy logic is a form of artificial intelligence software and it may be considered as a subset of artificial intelligence. Artificial intelligence and EI are two sides of the same coin. One side of the coin there are skills and on the other side there are advanced technological competence. Fuzzy logic controllers [24] are the mechanics on which fuzzy rule base system works like human thinking. Due to its logical reasoning ability [25] many researches [26, 27] have also introduced controller based on fuzzy logic. Fuzzy logic-based EI system has also been applied to test the emotional literacy of students [28]. Due to the assumption of truth value only, fuzzy logic fails to deal many vague, hesitation, and doubtful cases. To eliminate this drawback, Atanassov [29] provided a study based on intuitionistic fuzzy logic. This study considers both the cases that are true and false. The sum of these two values together with hesitation margin is always equal to 1. Further, the truth value represents the membership grade; false value indicates the non-membership grade, respectively. There are so many applications of intuitionistic fuzzy logic in various fields [30, 31]. But in the present era of contradiction, both the logics, i.e., traditional fuzzy logic and intuitionistic fuzzy logic fails to handle the situations, where we have contradictory, non-contradictory, and conflicting factors. Montiel et al. [32] gave a mediate solution to such problems called the mediative fuzzy logic. It may apply as an extension of previously existing logics (i.e., traditional and intuitionistic fuzzy logics). Many interesting applications of mediative fuzzy logic have also been studied [33–35]. To deal with conflicting factors, we will use the mediative fuzzy logic due to its nature to deal with conflicting factors gave the solution in mediative forms.

In this present work, we will develop a mediative fuzzy logic-based emotional inference system in the management of the pandemic days based on three obstacles, i.e., knowledge perception, conflicting factors, and EI. We take five factors as an input for the system namely, self-awareness, self-management, relationship, self-motivation, and empathy. And the output based on mediative fuzzy logic for emotional intelligent system has categorized into five linguistic variables namely, financial crisis, psychological behavior, interpersonal effects (I) and clinical implication for health issues and education loss.

4.2 BASIC CONCEPTS

4.2.1 FUZZY LOGIC

When uncertainty, fuzziness, vagueness, and impreciseness exist in the experimental data or real-life problems, then fuzzy logic (given by Zadeh in 1965 [23]) is used to deal with these aspects. It is a computational based paradigm on human being thinking.

4.2.2 INTUITIONISTIC FUZZY LOGIC

If fuzzy logic fails to deal with the state where an unfavorable or antagonistic case exists, then intuitionistic fuzzy logic is used to deal with this situation. In this logic, we deal with favorable as well as unfavorable cases together including the hesitation part.

Emotional Intelligence and Its Management

4.2.3 MEDIATIVE FUZZY LOGIC

In the present scenario where EI, knowledge, and conflicting factors affect the human stay at home then intuitionistic fuzzy logic and traditional fuzzy logic are not adroit to deal with these situations, due to the involvement of three types of contradictory factors. To handle such kind of contradictory, non-contradictory, and doubtful situation, Montiel et al. in 2005 [32] introduced the concept of mediative fuzzy logic. In this logic, we try to find out a mediate solution to the problem at the optimal degree of acceptance with the help of existing contradictory factors. Contradictory factor works as a minimum of agreement or disagreement factor.

4.2.4 FUZZY LOGIC CONTROLLER FOR RULE-BASED SYSTEM

To perform the judgment of a human being, we need a paradigm that can easily deal with conflicting situations to form a self-operating program so that one can have an ability to perform as a human being. Fuzzy logic controller has the ability to handle such type of circumstances there are three types of fuzzy logic controller have been studied, namely, Mamdani, Sugeno, and Tsukamoto. In this work, we use Mamdani-type fuzzy logic controller. Further, we extended the inference system with the help of mediative fuzzy logic. For the proposed approach of this work, we need to define the membership and non-membership values for the factors and then we define the fired fuzzy rules.

4.3 MEMBERSHIP AND NON-MEMBERSHIP FUNCTIONS FOR THE FACTORS

In this proposed work, we used the concept of self-assessment tool introduced by Cartwright and Solloway [36], which was adapted from Goleman's model of EI [37] to measure the conflict at home during a pandemic. A framework on mediative fuzzy-based EI is developed to calculate various linguistic categories of output for the measurement of EI. The proposed framework gave the most appropriate output designed to raise their EI effectively. We constructed the membership and non-membership functions for the input and output factors that have been included in our study by generating the codes in MATLAB. These all are represented by Figures 4.1–4.11.

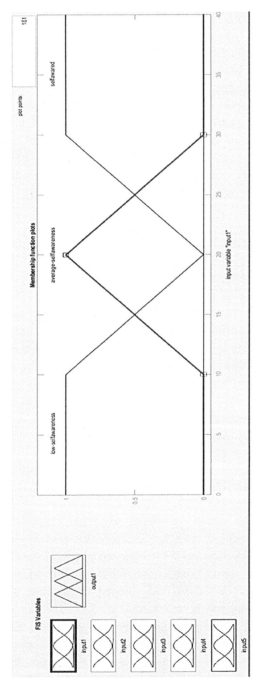

FIGURE 4.1 Membership function for self-awareness.

Emotional Intelligence and Its Management 59

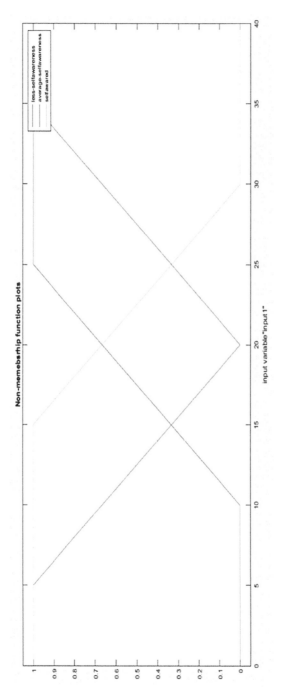

FIGURE 4.2 Non-membership function for self-awareness.

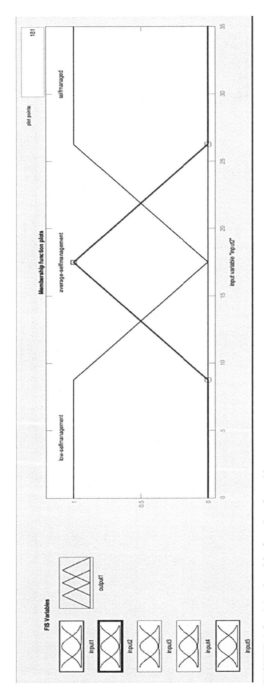

FIGURE 4.3 Membership function for self-management

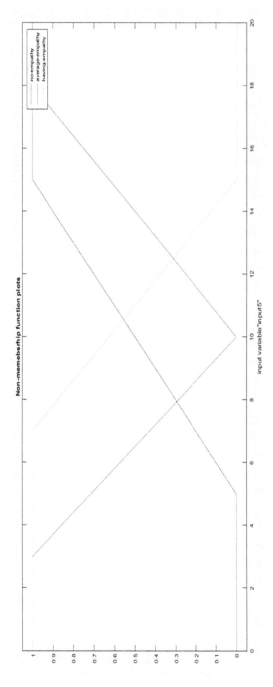

FIGURE 4.4 Non-membership function for self-management.

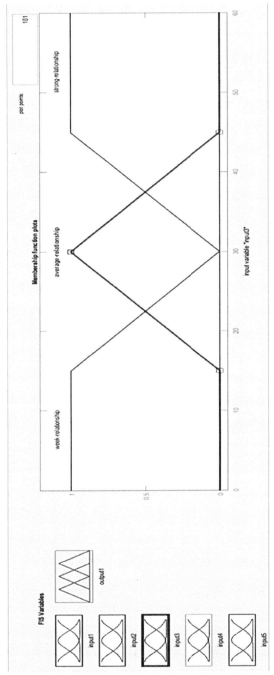

FIGURE 4.5 Membership function for relationship.

Emotional Intelligence and Its Management 63

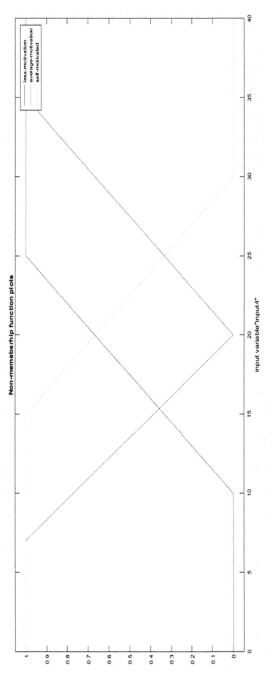

FIGURE 4.6 Non-membership function for relationship.

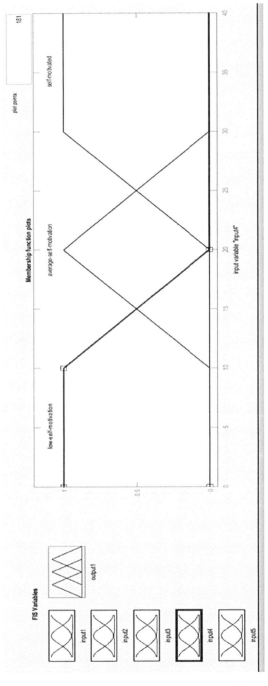

FIGURE 4.7 Membership function for self-motivation.

Emotional Intelligence and Its Management 65

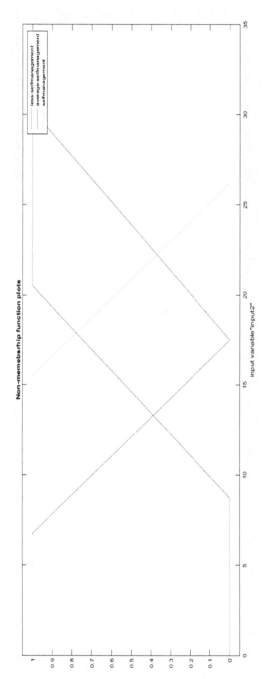

FIGURE 4.8 Non-membership function for self-motivation.

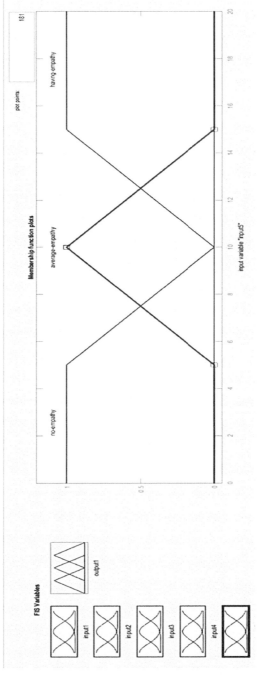

FIGURE 4.9 Membership function for empathy.

Emotional Intelligence and Its Management 67

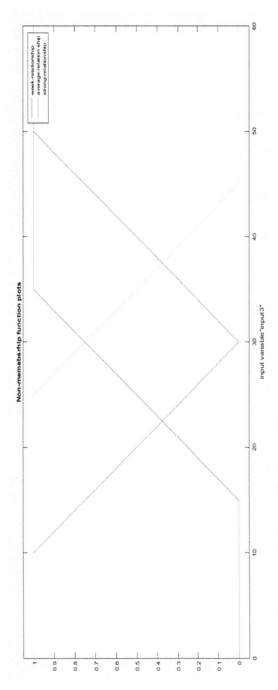

FIGURE 4.10 Non-membership function for empathy.

68 Emotional Intelligence for Leadership Effectiveness

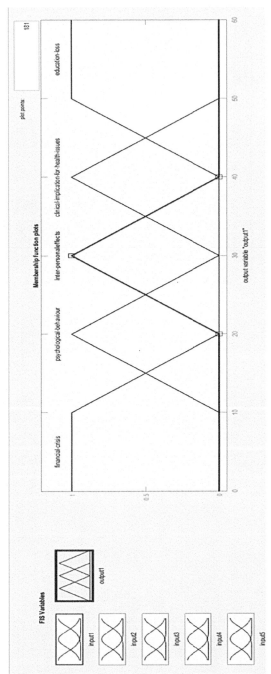

FIGURE 4.11 Membership function for output variables.

Emotional Intelligence and Its Management 69

Mathematically, the numerical categorization of the factors into the membership and non-membership functions for each input factors with their three linguistic categories are shown in Table 4.1.

TABLE 4.1 Input Factors and Their Corresponding Membership and Non-Membership Grades

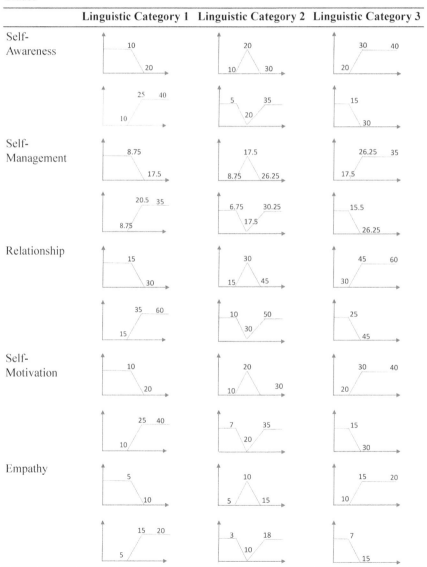

For the non-membership value of the output factor, the ranges of the linguistic variables are lies between:

- [0, 28] (for F);
- [5, 28] (for P);
- [15–38] (for I);
- [25–48] (for C); and
- [35–60] (for E).

4.4 FUZZY INFERENCE SYSTEM FOR EMOTIONAL INTELLIGENCE (EI) MODEL

We developed the fuzzy IF-THEN rules for the mediative fuzzy logic based emotional intelligent system according to the requirement and the nature of the particular problem. From the inference of these fuzzy rules, we will get a comprehensive mediate value. In this work, we have taken the 10 fired fuzzy rules. The program for these fuzzy rules is given in Table 4.2 and these rules are shown in Table 4.3.

The rules defined in the logic-based inference system can be summarized in Table 4.3.

4.5 NUMERICAL COMPUTATIONS

The numerical computations for the proposed model of EI based on the mediative fuzzy logic have been carried out with the MATLAB software for the conflicting factors. The significance of the proposed model and individual conflicting factor and the lack of hesitational part are tested using the input as well as the output factors. A comparison between the intuitionistic and mediative logic based out is also obtained. We have taken different inputs and calculated the numerical values which have been shown in Tables 4.4–4.6.

4.6 CONCLUSION

This work represents a novel framework to model and process EI skills in the context of conflict at home during pandemic period. The proposed

Emotional Intelligence and Its Management 71

TABLE 4.2 Mediative Logic-based Inference System for the Fired Fuzzy Rules of the Proposed Model

	IF					THEN
	Input 1 (self-awareness)	Input 2 (self-management)	Input 3 (relationship)	Input 4 (self-motivation)	Input 5 (Empathy)	Output
1.	low	low	week	low	no	Psychological-behaviour
2.	low	low	strong	low	having	Financial-crisis
3.	average	average	average	Low	average	Inter-personal effects
4.	Self	average	average	average	no	Clinical implication for health issues
5.	low	low	week	low	having	Education loss
6.	average	average	average	average	average	Financial crisis
7.	self	self	week	low	average	Psychological-behaviour
8.	low	low	strong	self	no	Psychological-behaviour
9.	average	self	week	low	no	Clinical implication for health issues
10.	self	self	strong	self	having	Education loss

TABLE 4.3 Summarized Fuzzy Rules

1.	If (input1 is low-self-awareness) and (input1 is low-self-management) and (input1 is week-relationship) and (input1 is low-self-motivation) and (input1 is no-empathy) then (output1 is psychological-behavior)
2.	If (input1 is low-self-awareness) and (input1 is low-self-management) and (input1 is strong-relationship) and (input1 is low-self-motivation) and (input1 is having-empathy) then (output1 is financial-crisis)
3.	If(input1 is average-self-awareness) and (input1 is average-self-management) and (input1 is average-relationship) and (input1 is low-self-motivation) and (input1 is average-empathy) then (output1 is inter-personal-effects)
4.	If (input1 is self-awared) and (input1 is average-self-management) and (input1 is average-relationship) and (input1 is average-self-motivation) and (input1 is no-empathy) then (output1 is clinical-implication-for-health-issues)
5.	If (input1 is low-self-awareness) and (input1 is low-self-management) and (input1 is week-relationship) and (input1 is low-self-motivation) and (input1 is having-empathy) then (output1 education-loss)
6.	If (input1 is average-self-awareness) and (input1 is average-self-management) and (input1 is average-relationship) and (input1 is average-self-motivation) and (input1 is average-empathy) then (output1 is financial-crisis)
7.	If (input1 is self-awared) and (input1 is self-managed) and (input1 is week-relationship) and (input1 is low-self-motivation) and (input1 is average-empathy) then (output1 is psychological-behavior)
8.	If (input1 is low-self-awareness) and (input1 is low-self-management) and (input1 is strong-relationship) and (input1 is self-motivated) and (input1 is no-empathy) then (output1 is psychological-behavior)
9.	If (input1 is average-self-awareness) and (input1 is self-managed) and (input1 is week-relationship) and (input1 is low-self-motivation) and (input1 is no-empathy) then (output1 is clinical-implication-for-health-issues)
10.	If (input1 is self-awared) and (input1 is self-managed) and (input1 is strong-relationship) and (input1 is self-motivated) and (input1 is having-empathy) then (output1 is education-loss)

TABLE 4.4 Numerical Computation I for the 10 Fired Fuzzy Rules Indicated by Rows

Input 1	Input 2	Input 3	Input 4	Input 5	Output 1	Intuitionistic-Based Output	Mediative-Based Output
15	15	20	15	8			
0.5, 0.34	0.28, 0.53	.67.25	.5.34	.4.3	P(0.28.25)	$Y1 = (1 − 0.47)\,20 + (0.47)\,20.296875 = 20.1395$	$Y1 = (1 − 0.47 − 0.125)\,20 +$ $(0.47 + 0.125)\,20.2968 = 20.5764$
0.5, 0.34	0.28, 0.53	0, 1	.5.34	0, 0.87	F(0.34)	$Y2 = (0.66)\,7.6357 = 5.039562$	$Y2 = (0.66)\,7.6357 = 5.039562$
0.5.34	.71.23	0.34.5	.5.34	.6.28	I(0.34.23)	$Y3 = (1−0.43)\,30 + (0.43)$ $30.23 = 30.0989$	$Y3 = (1 − 0.43−0.115)\,30 + (0.43$ $−0.115)\,30.23 = 30.12535$
0, 1	.71.23	0.34.5	.5.38	.4.3	C(0.23)	$Y4 = (0.77)\,40.213636 = 30.9645$	$Y4 = (0.77)\,40.213636 = 30.9645$
0, 1	.28, 0.53	.67.25	.5.34	.4.3	E(0.23)	$Y5 = (0.77)\,54.0996$	$Y5 = (0.77)\,54.0996$
						41.6567	41.6567
.5.34	0.1	0.34.5	.5.38	.6.28	F(0.25)	$Y6 = (0.75)5.5154$ $= 4.1365$	$Y6 = (0.75)5.5154$ $= 4.1365$
0, 1	0.1	0.34.5	.5.34	.6.28	P(0.28)	$Y7 = (0.72)20.28$ $= 14.6016$	$Y7 = (0.72)20.28$ $= 14.6016$
0.5, 0.34	.28, 0.53	0.1	0.1	.4.3	P(0.3)	$Y8 = (0.7)20.3 = 14.21$	$Y8 = (0.7)20.3 = 14.2$
0, 1	0.1	.67.25	.5.34	.4.3	C(0.25)	$Y9 = (0.75)40.25 = 30.1875$	$Y9 = (0.75)40.25 = 30.1875$
0, 1	0.1	0.1	0.1	0, 0.87	E(0.87)	$Y10 = (0.13)53.762798 = 6.9591$	$Y10 = (0.13)53.762798 = 6.9891$
						Aggregated = 19.8023862	**Aggregated = 19.8487**

TABLE 4.5 Numerical Computation II for Different Inputs and for the 10 Fired Fuzzy Rules Indicated by Rows

Input 1	Input 2	Input 3	Input 4	Input 5	Output 1	Intuitionistic	Mediative
25	25	20	10	12			
0.5, 0.34	0, 1	.67.25	5.34	0.7	P (0.25)	Y1 = (0.75)20.296875 = 5.0742	Y1 = (0.75)20.296875 = 5.0742
0.5, 0.34	0, 1	0, 1	.5.34	.4.37	F (0.34)	Y2 = (0.66)7.6357 = 5.039563	Y2 = (0.66)7.6357 = 5.039562
0.5.34	.14.23	0.34.5	.5.34	.6.25	I (0.14.23)	Y3 = (1−0.63)30+(0.63) 30.23 = 30.1449	Y3 = (1−0.63−0.07)30+(0.7) 30.23 = 30.161
0, 1	.14.23	0.34.5	.5.38	0.7	C (0.23)	Y4 = (0.77)40.213636 = 30.9645	Y4 = (0.77)40.213636 = 30.9645
0, 1	0, 1	.67.25	.5.34	.4.37	E (0.25)	Y5 = (0.75)54.0131 = 40.5098	Y5 = (0.75)54.0131 = 40.5098
.5.34	.14.23	0.34.5	.5.38	.6.25	F (0.14.23)	Y6 = (1−0.63)7.78+(0.63) 5.4726 = 6.3263	Y6 = (1−0.63−0.07)7.78+(0.7) 5.4726 = 6.16482
0, 1	.14.23	0.34.5	.5.34	.6.25	P (0.23)	Y7 = (0.77)20.23 = 15.5771	Y7 = (0.77)20.23 = 15.5771
0.5, 0.34	0, 1	0, 1	0, 1	0.7	P (0.34)	Y8 = (0.66)20.34 = 13.4244	Y8 = (0.66)20.34 = 13.4244
0, 1	.14.23	.67.25	.5.34	0.7	C (0.23)	Y9 = (0.77)40.213636 = 30.9645	Y9 = (0.77)40.213636 = 30.9645
0, 1	.14.23	0, 1	0, 1	.4.37	E (0.23)	Y10 = (0.77)54.0996 = 41.6567	Y10 = (0.77)54.0996 = 41.6567
						= 21.9682	**= 21.9536**

Emotional Intelligence and Its Management 75

TABLE 4.6 Numerical Computation III for Different Inputs and for the 10 Fired Fuzzy Rules Indicated by Rows

Input 1	Input 2	Input 3	Input 4	Input 5	Output 1	Intuitionistic	Mediative
28	15	20	15	8			
0. 1	.28, 0.53	.67.25	.5.34	.4.3	P (0.25)	$Y1 = (0.75)20.296875$ 5.0742	$Y1 = (0.75)20.296875$ 5.0742
0. 1	.28. 0.53	0. 1	.5.34	0, 0.87	F (0.34)	$Y2 = (0.66)7.6357$ $= 5.039562$	$Y2 = (0.66)7.6357$ $= 5.039562$
.2.14	.71.23	0.34.5	.5.34	.6.28	I (0.2.23)	$Y3 = (1-0.57)30+(0.57)30$ $= 1$	$Y3 = (1-0.57-0.1)30+(0.67)30$ $= 1$
.8.14	.71.23	0.34.5	.5.38	.4.3	C (0.34.23)	$Y4 = (1-0.43)40+(0.43)39.79$	$Y4 = (1-0.43-0.115)$ $40+ (0.43+.115)39.79$
0. 1	.28, 0.53	.67.25	.5.34	.4.3	E (0.25)	$Y5 = (0.75)54.0131$ 40.5098	$Y5 = (0.75)54.0131$ 40.5098
.2.14	0. 1	0.34.5	.5.38	.6.28	F (0.28)	$Y6 = (0.75)5.5154$ $= 4.1365$	$Y6 = (0.75)5.5154$ $= 4.1365$
.8.14	0. 1	0.34.5	.5.34	.6.28	P (0.28)	$Y7 = (0.72)20.28$ 14.6016	$Y7 = (0.72)20.28$ $= 14.6016$
0. 1	.28, 0.53	0. 1	0. 1	.4.3	P (0.3)	$Y8 = (0.7)20.3$ $= 14.21$	$Y8 = (0.7)20.3$ $= 14.21$
.2.14	0. 1	.67.25	.5.34	.4.3	C (0.25)	$Y9 = (0.75)40.25$ $= 30.1875$	$Y9 = (0.75)40.25$ $= 30.1875$
.8.14	0. 1	0. 1	0. 1	0, 0.87	E (0.87)	$Y10 = 41.6567$ $= \mathbf{20.45}$	$Y10 = 41.6567$ $= \mathbf{20.42}$

model is based on mediative fuzzy logic. Five areas of EI competencies that help in identify the conflict. Mediative fuzzy linguistic variables and mediative fuzzy logic are used to find out the appropriate uncertainties as well as the complexities of modeling the EQ skills and their assessment to get the output. The mediative fuzzy-based system is expected to provide a suitable tool for the better orientation of their conflictions factors that help to succeed in their personal lives. The entire work done in this chapter illustrates the following points:

- This work presented a new approach to evaluate the conflict at home during pandemic. The proposed approach plays a vital role in nurturing and to develop the EQ skills level to overcome the conflicts and learning perspective.
- A mediative fuzzy logic-based EI system and processing its framework is developed for the comprehensive model and to capture uncertainties and contradiction in the proposed framework. This approach is expected to provide as much as accuracy to understand the statements of conflict at home during a pandemic. The fuzzy linguistic variables and mediative fuzzy logic is used to handle the contradiction and complexities of the modeling of the EQ skills and their assessment in the results.
- The basic objective of manual calculation is to obtain an appropriate output for the EI. The desired output cannot be calculated by using any software or simulation-based technique.
- In case of intelligent quotient any program may use in a proper way to get each solution. But in this case, a program-based computation gives the best optimal result. But here, the numerical computation is performed by manual calculation as well as the software-based calculations because in the case of EQ, the calculation cannot be measured exactly by machine or any simulation technique, so such condition can easily be calculated manually.
- In these computational (Tables 4.4–4.6) the output of mediative logic-based approach is more optimal than the intuitionistic fuzzy logic-based approach. Due to the consideration of contradictory factor, the obtained output is quite better than the intuitionistic logic-based system output. Numerical computation I for Table 4.4, indicates that the conflict may arise at home during the pandemic arises due to the psychological behavior of the person or financial crisis. In the case of numeric computation II and III in Tables 4.5 and 4.6,

Emotional Intelligence and Its Management 77

conflict arises due to some interpersonal effects and psychological behaviors of the person.

- To compute the traditional fuzzy output, we applied the MATLAB [38] software, and by using this we gained the output of fuzzy logic-based system. And for further study, we used the manual computation to get the approximated solution of the problem.
- A comparative study between intuitionistic fuzzy logic and mediative is done in this work. The superiority of the mediative fuzzy logic over intuitionistic fuzzy enhanced in this work. With the consideration of mediative fuzzy logic conflict at home during pandemic can easily be handled.

KEYWORDS

- **conflicting factors**
- **education loss**
- **emotional intelligence**
- **intelligence quotient**
- **intuitionistic fuzzy logic**
- **MATLAB**
- **mediative fuzzy logic**

REFERENCES

1. Bond, B. B., & Manser, R., (2009). *Emotional Intelligence Interventions to Increase Student Success.* Toronto: Higher Education Quality Council of Ontario.
2. Coleman, A., (2008). *A Dictionary of Psychology* (3rd edn., p. 248). Oxford: Oxford University Press.
3. Austin, W., Evans, P., Goldwater, R., & Potter, V., (2005). A preliminary study of emotional intelligence, empathy and exam performance in first year medical students. *Personality and Individual Differences, 39,* 1395–405.
4. Ahmetoglu, G., Leutner, F., & Chamorro-Premuzic, T., (2011). *EQ-Nomics: Understanding the Relationship Between Individual* differences in trait emotional intelligence and entrepreneurship. *Personality and Individual Differences, 51*(8), 1028–1033.

5. Blattner, J., & Bacigalupo, A., (2007). Using emotional intelligence to develop executive leadership and team and organizational development. *Consulting Psychology Journal: Practice and Research, 59,* 209–219.
6. Bratton, V. K., Dodd, N. G., & Brown, F., (2011). The impact of emotional intelligence on accuracy of self-awareness and leadership performance. *Leadership & Organization Development Journal, 32,* 127–149.
7. Baba, M. M., (2020). Navigating COVID-19 with emotional intelligence. *International Journal of Social Psychiatry, 66.*
8. Antonakis, J., Ashkanasy, N. M., & Dasborough, M. T., (2009). Does leadership need emotional intelligence? *The Leadership Quarterly, 20,* 247–261.
9. Gorkan, A. G., Leutner, F. K., & Premuzic, T. C., (2011). EQ-nomics: Understanding the relationship between individual differences in trait emotional intelligence and entrepreneurship. *Personality and Individual Differences, 51,* 1028–1033.
10. Nelis, D., Quoidbach, J., Mikolajczak, M., & Hansenne, M., (2009). Increasing emotional intelligence: (How) is it possible?. *Personality and Individual Differences, 47,* 36–41.
11. Francesca, M. D., (2013). *Want an MBA From Yale? You're Going to Need Emotional Intelligence* (p. 15). Bloomberg Business Week.
12. Pool, L., & Qualter, P., (2012). Improving emotional intelligence and emotional self-efficacy through a teaching intervention for university students. *Learning and Individual Differences, 22,* 306–312.
13. Poornima, A., & Sujatha, S., (2020). Emotional intelligence and leadership effectiveness: Mediation effect of trust and organizational citizenship behavior of the middle-level managers in life insurance corporations in Chittoor district –a sequential mediation with two mediators. *Journal of Critical Reviews, 16,* 850–858.
14. McCrae, R., (2000). Emotional intelligence from the perspective of the five-factor model of personality. In: Bar-On, R., & Parker, J. D. A., (eds.), *The Handbook of Emotional LODJ, Intelligence: Theory, Development, Assessment, and Application at Home, School and in the Workplace* (pp. 263–276). Jossey-Bass/Wiley, New York, NY.
15. Mayer, J. D., Caruso, D. R., & Salovey, P., (2000).c Selecting a measure of emotional intelligence: The case for ability scales. In: Bar-On, R., & Parker, J. D. A., (eds.), *The Handbook of Emotional Intelligence: Theory, Development, Assessment, and Application at Home, School and in the Workplace.* Jossey-Bass/Wiley, New York, NY.
16. McColl-Kennedy, J. R., & Anderson, R. D., (2002). Impact of leadership style and emotions on subordinate performance. *The Leadership Quarterly, 13,* 545–599.
17. Roberts, R. D., Zeidner, M., & Matthews, G., (2001). Does emotional intelligence meet traditional standards for intelligence?. *Emotion, 1,* 196–231.
18. Wong, C. S., & Law, K. S., (2012). The effects of leader and follower emotional intelligence on performance and attitude: An exploratory study. *The Leadership Quarterly, 13,* 243–274.
19. Vierimaa, J. C., (2013). *Emotional Intelligence and Project Leadership: An Explorative Study.* Master's thesis.
20. Kerr, R., & Garvin, J., (2006). Norma Heaton and Emily Boyle emotional intelligence and leadership effectiveness. *EI and Leadership Effectiveness, 27,* 265–279.

Emotional Intelligence and Its Management 79

21. Prati, L. M., Douglas, C., Ferris, G. R., Ammeter, A. P., & Buckley, R. M., (2003). Emotional intelligence, leadership effectiveness, and team outcomes. *The International Journal of Organizational Analysis, 11*, 21–40.
22. Jordan, P. J., James, C. A., & Ashkanasy, N. M., (2006). Evaluating the claims: Emotional intelligence in the workplace. Evaluating the claims. In: Murphy, K. R., (ed.) *A Critique of Emotional Intelligence: What Are the Problems and How Can They be Fixed* (pp. 198–210). Mahwah, NJ: Lawrence Erlbaum Associates.
23. Zadeh, L. A., (1965). Fuzzy sets. *Information and Control, 8*, 338–356.
24. Zadeh, L. A., (1975). The concept of a linguistic variable and its application to approximate. *Information Sciences, 8*, 301–357.
25. Kazuo, T., & Wang, H. O., (2001). *Fuzzy Control Systems Design and Analysis: A Linear Matrix Inequality Approach.* John Wiley and Sons Publication.
26. Dhiman, N., & Sharma, M. K., (2020). Fuzzy logic inference system for identification and prevention of coronavirus (COVID-19). *International Journal of Innovative Technology and Exploring Engineering (IJITEE), 9*, 1575–1580.
27. Mamdani, E. H., (1974). Applications of fuzzy algorithms for control of a simple dynamic plant. *Proc. IEE, 121*, 1585–1588.
28. Bouslama, F., Housley, M., & Steele, A., (2015). Using a fuzzy logic-based emotional intelligence framework for testing emotional literacy of students in an outcomes-based educational system. *Journal of Network and Innovative Computing, 3*, 105–114.
29. Atanassov, K. T., (1986). Intuitionistic fuzzy sets. *Fuzzy Sets and Systems, 20*, 87–96.
30. Atanassov, K. T., (1989). More on intuitionistic fuzzy sets. *Fuzzy Sets and Systems, 33*, 37–45.
31. Atanassov, K. T., (1983). Intuitionistic fuzzy sets. In: Sgurev, V., (ed.), *VII ITKR'S Session, Sofia.*
32. Montiel, O., Castillo, O., Melin, P., Rodríguez, D. A., & Sepúlveda, R., (2005a). Reducing the cycling problem in evolutionary algorithms. In: *Proceedings of ICAI-2005* (pp. 426–432).
33. Montiel, O., & Castillo, O., (2007). Mediative fuzzy logic: A New approach for contradictory knowledge management. *European Journal of Operation Research, 179*, 220–233.
34. Dhiman, N., & Sharma, M. K., (2019). Mediative sugeno's-TSK fuzzy logic-based screening analysis to diagnosis of heart disease. *Applied Mathematics, 10*, 448–467.
35. Dhiman, N., & Sharma, M. K., (2019). Mediative fuzzy logic approach based on sugeno's-TSK model for the diagnosis of diabetes. *24th International Conference of CONIAPS in India.*
36. Cartwright, A., & Solloway, A., (2007). *Emotional Intelligence: Activities for Developing you and Your Business.* Taylor & Francis.
37. Goleman, D., (1996). *Emotional Intelligence: Why It Can Matter More than IQ.* New York: Bantam.
38. Mathworks, (2022). *MATLAB and the Fuzzy Logic Toolbox* [Online]. Available: www.mathworks.com (accessed on 5 July 2022).

CHAPTER 5

Unlocking the Mask: A Look at the Role of Leadership and Innovation Amid the COVID-19 Pandemic Crisis

SABZAR AHMAD PEERZADAH, SABIYA MUFTI, and SHAYISTA MAJEED

School of Business Studies University of Kashmir, Srinagar, Jammu and Kashmir–190006, India

ABSTRACT

The outbreak of COVID-19 has the hallmarks of a "landscape-scale" crisis: an unexpected event or sequence of events of enormous scale and overwhelming speed, resulting in a high degree of uncertainty that gives rise to disorientation, a feeling of lost control, and strong emotional disturbance. COVID-19 has made its impact on almost every area of life, from personal to professional today more than ever before. The humanitarian toll taken by COVID-19 creates fear among employees and other stakeholders. However, the word "crisis" is composed of two characters in the Chinese language: one representing a threat, the other representing opportunity. There is no question that as we look back on the ongoing health crisis, we will learn that it has resulted in many innovations: new medicines and medical devices, enhanced healthcare processes, breakthroughs in the production and supply chain, and new collaboration techniques. What these innovations will have in common is that problems will be solved, which is always at the core of innovation.

Emotional Intelligence for Leadership Effectiveness: Management Opportunities and Challenges During Times of Crisis. Mubashir Majid Baba, Chitra Krishnan, & Fatma Nasser Al-Harthy (Eds.)
© 2023 Apple Academic Press, Inc. Co-published with CRC Press (Taylor & Francis)

In a time when an organization is in crisis, what role leaders play, and how their programs can generate value is of utmost importance. Because innovation and leaders play important roles in helping the organization succeed and drive development within this new climate. The leaders can sink an organization faster than a ship with a leak or raise an organization from the ashes. When all is smooth sailing, the true test of leadership does not happen. Instead, leadership during a crisis is tested. Moreover, seeing the possibilities coming from this crisis is not like being able to seize them. Therefore, considering this context, the current chapter illustrates the role played by leadership during the pandemic crisis, the need for innovation in the crisis, and the measures that leaders may take to advance innovation in the crisis to mitigate problems, followed by a discussion and conclusion of the chapter.

5.1 INTRODUCTION

The nCoV 2019 (COVID-19) is a new category of coronavirus which has not been witnessed till date. The World Health Organization (WHO) professed COVID-19 a pandemic on March 11, 2020 and graded it as a "CoV" virus, which is "a large group of viruses that cause illness extending from the common cold to more serious diseases like middle east respiratory syndrome (MERS-CoV) and severe acute respiratory syndrome (SARS-CoV). COVID-19's ongoing uncertainty has raised a huge threat to the healthcare industry in particular, as well as other industries in general" [1]. Almost immediately, universities, colleges, and schools all over the world shut their doors and changed their teaching from face-to-face to the internet. Staff and students have reacted with incredible flexibility, revising delivery and evaluation procedures which otherwise could have elongated to several months to complete via conventional channels. Organizations in any industry and location are experiencing unparalleled intensity and magnitude of transition. Manufacturing companies have shifted their activities to accommodate the strong and immediate demand for critical goods such as ventilators, face masks, hand sanitizers, and paracetamol. Governments also developed comprehensive strategies to assist persons and organizations facing bankruptcy or layoffs putting economic issues aside to concentrate on societal goals as for instance, safeguarding the interests of the needy, aiding people in financial need, and improving key

Unlocking the Mask: A Look at the Role of Leadership 83

public services (particularly health and social care). Communities have come together in ways that have not been seen since World War II, giving encouragement and reassurance to the aged and lonely, compromising personal liberty for the greater good, and innovating new ways to interact, communicate, and collaborate.

Though it is not unusual to bracket "leadership" with the characteristics and behaviors of human "leaders," the COVID-19 pandemic highlights the importance of individuals and communities working together to accomplish leadership outcomes. However, many national figures have failed to comprehend the scope and complexities of the challenges raised by COVID-19. It is worth mentioning here that both local leadership and administration have proven much more successful in mobilizing governmental, corporate, non-profit, and local associations and organizations to unite and respond. Place-based leadership responds to the environment in which it operates, bringing together various viewpoints and experience to resolve matters of interest to residents of a specific area, which will be critical not only in coping with the immediate consequences of COVID-19 but also in the lengthy process of reconstruction and regeneration that will accompany the pandemic [2]. Crises have proven to be catalysts for innovation in the past. The way we used to live got drastically transformed due to the COVID-19 pandemic, thereby defying any question that our society will be changed fundamentally after the threat has passed. However, If the pain is never the same, most global crises cause considerable economic disruption and can rip societies apart but they still motivate innovation. Many workplaces and educational establishments have had to rapidly adjust to working remotely that was quite challenging as well. However, the current crisis is due to a lack of foresight and the ability to act on that foresight, rather than a lack of ingenuity or adaptability. Experts have cautioned that due to climate disruptions, population growth, and amplified global interconnectedness, such type of pandemics can occur frequently. Institute for the Future (IFTF) in the USA, developed forecasts based on bio-disasters more than 10 years ago, focusing on zoonotic diseases (those transferred from animals to humans), anticipatory quarantines, and other viral risks. The Institute developed Superstruct, a large-scale participatory game in which thousands of people worked to plan and execute solutions to five "super-threats," including a pandemic, 10 years before the current outbreak. We were not the only ones who stressed the importance of being prepared. Many analysts, including those in government, have raised the

84   *Emotional Intelligence for Leadership Effectiveness*

alarm. However, there are few incentives integrated into our systems for people to gain knowledge about the future systemically and then act accordingly. Short-term results are rewarded in our capital markets and businesses; politicians are concerned with the next election, and most of the media relies on news stories rather than in-depth research. Hence, in such a circumstance, the responsibility lies with top leadership and the innovative solutions to the problems caused by the pandemic. The next section of the chapter elaborates the role played by the top leadership at this turbulent time, followed by the significant role of innovation, the measures needed to be taken by leadership to advance innovation. Subsequently, the discussion and conclusion sections of the chapter are presented.

5.2   LEADERSHIP AMID PANDEMIC CRISIS

Crises can be triggered by disasters, pandemics, and emergencies, amongst other woes. While crises differ in terms of scope, their predisposition to become global has been on the rise since the past two decades, with people getting more connected globally. The dawn of globalization with its upturn in international travel and the progression of the technology sector has conferred the world with gigantic opportunities and threats alongside. During December 2019, a novel coronavirus (SARS-CoV-2) was spotted in Wuhan, China, and soon its rate of infection leading to fatality was on the surge. The fatalities owed to a complex respiratory syndrome named COVID-19. Just within a short period of three months, the virus had spread to over 100 nations in the world and 219 by March 2021 [3, 4]. The aftermaths of the impact, whether negative or positive, depending on leadership and personal initiatives in the society were so obvious [5]. The crisis is something unavoidable since time immemorial and blatantly crisis' management is necessary for survival and prosperity of human life.

As per Zamoum and Gorpe: "What composes a crisis is not easily agreed upon, however, notwithstanding lack of clarity, there are specific conditions as given in extant literature. For instance, crisis has six characteristics which are rare, significant, high impact, ambiguous, urgent, and involve high stakes. Crisis involves a period of discontinuity, a situation where the core values of the organization/system are under threat, and this calls for critical decision making. There is a destabilizing effect of the organization and its stakeholders and escalation of one or more issues, errors or procedures are expected in this period [6]."

A global challenge such as the coronavirus outbreak requires a global response. Better coordination and collaboration among nations was highly required, however, we saw countries sealing off their borders as a precautionary measure to decelerate the reach of the pandemic, denunciating other nations and competing for scarce resources, and even engaging in absurd conspiracy theories. Responding to a grand challenge like a pandemic requires cross-country and cross-sector collaboration (e.g., partnerships with NGOs, public sector entities, and even competitors). While providing a simple description of what organizations need to endure a pandemic, Nohria [7] underscores the worth of distributed leadership (divergent to centralized leadership), dispersed workforce (as opposed to the concentrated workforce), and networked structure (as opposed to hierarchical structure), and pointing to the need for a global network of such people in organizations who can align and adapt as events evolve, countering instantly and appropriately to any disruptions. He also highlights the importance of global alliances, suggesting that companies should aim at co-developing adequate crisis responses with partners and even competitors.

The bottom line of all this is simple: Local self-isolation and social distancing may be adequate measures to restrain the reach of the virus from an epidemiological perspective. In global politics and business, these are a recipe for disaster.

The value of leadership in times of emergency is widely recognized. This pandemic has brought to the forefront the deep failures of our leaders to work together, and this time the outcome of these failures can be truly devastating. Our leaders are failing us once again. Subsequent to SARS, Ebola, the 2008–2009 worldwide economic recession, and other pivotal events, our governments cannot and will not function in a composed and synergetic manner toward a common goal. That common goal now is to defeat probably the most serious worldwide peril to our civilization that is unparalleled. Any worldwide problem of this description, significance, and scale positively requires a global approach [8]. Some glaring failures of leadership and accountability on the part of corporate executives (e.g., the Uber CEO's negation to bear responsibility for the health and safety of their workers in this crisis period). The COVID-19 pandemic has shed light on a significant and fatal lack of responsible global leadership. Looking for an explanation for the differences is a tricky business as testing regimes are different, so are infrastructure and intensive care coverage. Additionally, the ways and means undertaken in various countries are driven by a vastly different sense of urgency. Still, watching the developments in the UK and the US, one cannot help but notice one factor that exacerbates the threat posed by the COVID-19 virus, and that is bad leadership [9]. To put it more bluntly, President Trump and Prime Minister Johnson have endangered their countries and own peoples by being, well, hypocritical, and self-inflated narcissists [10] other than being responsible global leaders in crisis. Gladly, the British Prime Minister survived his infection, but it is telling that his health, and not his failures, blunders, and the fact that people are dying by the hundreds, became the dominant story [11]. In contrast, the calm, considerate, and caring approach were taken by Mrs. Merkel in Germany or the science-based, compassionate but decisive leadership of Jacinta Adern in New Zealand with clear communication and wide-spread testing and treatment has not only helped to save many lives but has worked in congenial ways with well-equipped, professional health systems under the experts' supervision toward a systemic response to the crisis.

Narcissism is not a crime, but it is a psychological disorder that can lead to devastating consequences in times of crisis when the world needs leaders who take charge, build teams of experts around them, consider their responses in light of the evidence and scientific advancements, communicate in a calm but compassionate way, with the greater public good in mind. Narcissistic leaders are said to hold an ostentatious sense of self-importance, daydreaming about infinite achievement, strength, and brilliance, feel they are "unique," demand undue praise, have unrealistic expectations of favorable treatment or automatic conformity with their expectations, and lack empathy [12]. In other words, they are rule-book narcissists. Mr. Johnson was picked for his office as a great communicator and tactician and eventually gave the UK its Brexit. But as the virus spread into Europe in February 2020, he went on a holiday with his fiancée somewhere in the British countryside. He only endorsed that the prevailing virus was the country's chief priority when the FTSE (Financial Times Stock Exchange) index went into freefall. But rather than taking a decisive action and coordinating the government's emergency response team, he took the weekend off, giving the virus three more days to run its course. He then started to entertain like the Netherlands the idea of authorizing the virus run its path to increase herd immunity, contrary to what experts have strongly advised. His performance since has been contradictory,

Unlocking the Mask: A Look at the Role of Leadership 87

indecisive, and out of tune, and as a consequence, the UK has lost precious time, and Mr. Johnson narrowly escaped his fatal infection. As for Mr. Trump, well he has done what a narcissistic leader would do: downplaying the severity of the looming pandemic (it is going to be just fine), blaming it on others (a Chinese virus), and when he could no longer overlook the developments attempting to hold the grandeur as a war-time president to fight the silent enemy. Further, he disarrayed numbers, made misleading or false claims regarding potential treatment, and Fox News and his allies on the religious right aided in circulating hazardous messages that COVID-19 was a hoax pretty much like climate change. Worse still, rather than uniting the country, he encouraged the state governors to compete for limited medical supplies and allowed an incoherent response to the crisis, allowing the virus to spread, which had devastating consequences in New York and specifically in the South, where mostly people are uninsured.

When the COVID-19 reached India the same year, the Indian leadership chose a path of blind myths and illogical deeds over a scientific direction. Although medical experts and scientists recommended a scientific solution to COVID-19, Indian leadership encouraged the masses to clang plates from balconies at 8 p.m. on March 22 to celebrate the efforts of health workers [13]. The plate clanging and clapping at 8 p.m. was done with the same pomp and performance that we see at baraat, jaagran, and Diwali. It would have been cute if the festivities took place on balconies, but instead, people stepped out of their homes and into the streets to come together' against the epidemic. And that negated the whole purpose of the exercise. Instead of honoring the real heroes, we finished disobeying precisely what they aspired for: be conscientious citizens who keep a social distance to prevent group transmission. Maybe the Prime Minister should have said it more explicitly for Indians. The issue is that the Prime Minister made no mention of the government's actual plans to combat the nCoV pandemic in India. What steps are being taken to test patients? Is there an adequate quantity of testing kits? How about the funding available for hospital rooms, nurses, beds, and ventilators? Are we going to follow the South Korean model with further testing and no lockdown to save the virus from spreading? Abstract proposals and a single day of semi-successful quarantine did not aid India, and that it is skeptical to deliver any assistance to India's top leadership. The world was still struggling with health and medical facilities before the pandemic; for instance, the Census [14] revealed that there were just 0.55 hospital beds per 1,000 inhabitants [14]. This adventure did not end here but it continued further to the candlelight show.

The countrymen were then encouraged to light a candle on April 5, to dispel the gloom brought on by the coronavirus. People were urged to put off electric lights for a short interval of nine minutes on that Sunday night and light torches, diyas, or flash mobile lights on their balconies or at doorsteps Just few minutes after the speech, social media platforms including Facebook, Instagram, and Twitter got overwhelmed with memes questioning the PM, "task kidhar hai?" (Where is the task?). When such types of questions are asked from the top leadership and they do not come up with an effective strategy, it depicts the lack of vision and planning to combat crisis on the part of leadership. What is clear is that from a responsible leadership perspective [15], most political and business leaders failed to amply tackle the global dimension of the prevailing crisis. In a crisis, leadership must be decisive, prudent yet caring, selfless, and oriented toward supporting people, with a consistent set of goals and a strong understanding of the systematic threats involved focused on facts and research, not on hunches, gut instincts, and self-serving agendas. The bad leadership of Mr. Johnson and Mr. Trump has tolled both nations, whose health care systems are ill-prepared, and consequently, too many people have died.

The pandemic did not stop here and so were the occurrences of failed leadership. As there were high speculations about its second wave that it could be more lethal than its predecessor, the present Indian government made certain decisions that were blatantly against the welfare of the common masses and consequently ensued huge disaster to human lives. To begin with, the central and state governments provided a nod to Kumb Mela 2021, a major pilgrimage for Hindus, for 30 days starting from April 1st to 30th. Though speculated as a superspreader of the COVID-19 virus, granting permission to such pilgrimage served as a sheer invite to the devastation. Apart from that, the election rallies held across various country parts where no COVID-19 SOPs (standard operating procedure) were followed added fuel to the fire. Here the political interest was given precedence over the national interest, thereby highlighting the failure of governance at all levels. As quoted by an editorial entitled India's COVID-19 emergency' in The Lancent, "In the face of warnings about the risks of superspreader events, the Indian government gave a green signal to religious festivals, luring millions of people from all around the nation, accompanied by huge political rallies-prominent for their lack of COVID-19 mitigation measures" [16].

Throughout the COVID-19 crisis, however, we have seen signs of politicians at all levels of government, industry, and civil society rising to the occasion, taking personal responsibility, and displaying genuine human interest. It was promising to see companies ranging from Alibaba to Amazon mobilize to aid in the pandemic's fight [17]. Via the Alibaba Foundation, for example, Jack Ma donated 6 million masks, more than 1 million testing kits, and 60,000 protective suits and face shields to various nations in Africa. When the entire world recuperates from this unparalleled crisis, we as scholars must examine leadership failure and emphasize the need for responsible global leaders [18], restating their qualities. We require such leaders who have a global mindset [19], who bear a sense of responsibility to all stakeholders, who pay attention to what others are saying and base their actions on a moral compass [20], and have a shared concern for the well-being of their constituencies and humanity as a whole [21]. Leaders, who are inclusive, kind, and considerate, and possess the vision to look at the bigger picture connecting past, present, and future as stewards of their countries and organizations. As the author [22] has put it recently, "let us practice responsible leadership ourselves by examining and advancing responsible leadership, along with other valuable topics, to contribute to the creation of a better world post-COVID-19."

5.3 INNOVATION AND THE PANDEMIC CRISIS

In the Chinese language, the term "crisis" basically comprises two characters: one that represents a threat and the other an opportunity. COVID-19 has made its impact on almost every area of life, from personal (how people sustain a life and work) to professional (their jobs) today more than ever before. Almost 90% of executives expect the COVID-19 fallout to radically change the way they conduct business over the next five odd years [23]. More than three-quarters of executives believe the recession would open up substantial new prospects for growth opportunities. However, this varies a lot from one sector to another. Organizations attempt to search for newer ways to achieve a competitive advantage in an uncertain and competitive world. Introducing innovations is one of these methods [24]. Innovation can be generally understood as the successful implementation/

execution of creative ideas within an organization [25]. Innovation is an opportunity, not a promise of success. Innovation is important not only in the medical and pharmaceutical fields [26] but also in all sectors of the economy in the COVID-19 period. Whilst a more stringent policy response to the pandemic was needed, businesses would eventually be influenced by it, with both short-term and long-term implications [27]. Firms in fast-changing technology markets that invest in R&D as a response to emerging global threats will be well placed to support not just their near-term survival, but also their potential innovative performance to remain competitive [28–34]. The Innovators have rushed in to help with the coronavirus pandemic response. Hand sanitizers have been adopted by breweries and distilleries all around the world. Valves for ventilators were rapidly manufactured using 3D printers by a start-up engineering company in Italy. Without a doubt, as we reflect on the current health crisis, we will discover that it has resulted in many innovations, including novel drugs and medical technologies, upgraded healthcare processes, production, and supply chain advances, and new collaboration strategies. These technologies will share one common feature: they will settle down issues, that is what innovation is all about. When things get harder, they will be driven by a deep human need to assist, to get connected with other people, and becoming a part of the solution.

Nevertheless, the generative nature of a crisis which paves the path for innovation is something more than simply an opportunity to sort out problems. Crisis offers unique challenges and innovators are compelled to think and act more freely to bring about swift and effective reform. Because of these requirements learning leaders encourage and empower employees to perform most of their innovative work in the best interest of their organizations. It seems quite impossible to know in exact terms whether or how the coronavirus spread will be slowed or stopped, or how we will be affected. But, if we use it carefully as an opportunity to innovate, learn, and evolve, we will realize that when it is over, we have given it our all and tried to do some good. Since "creativity and innovation love crises and restraints," the latest public health crisis is catalyzing innovation in organizations across the world [35]. Every organization has a record for bringing the most creative and solution-oriented technologies to market in a short time. Many factors must be considered to navigate the invention process to achieve a competitive advantage. The most critical element impacting creativity and innovation has been proposed to be leadership

[25]. Many factors affect creativity and innovation, but leaders and their actions have been said to have an especially strong influence [37]. Furthermore, leaders in innovation cannot focus on predefined structures; instead, they must be able to induce structure and provide guidance to work that lacks it [37]. Leaders can drown an organization faster than a ship with a leak or lift an organization from the ashes. Nevertheless, identifying the opportunities presented by this crisis is quite different from taking the advantage. Prioritizing innovation today can lead to unlocking post-crisis growth. There is no question that the global health crisis has resulted in a slew of innovations: innovative treatments and medical technologies, improved clinical processes, logistics, and supply chain breakthroughs, and creative collaborative strategies.

5.4 THE MEASURES THAT LEADERS NEED TO TAKE TO ADVANCE INNOVATION

The concept of the butterfly effect known from chaos theory illustrates the idea that small changes such as the movement of the wings of a butterfly can cause large-scale systemic change. In terms of the corona crisis, it was presumably not the metaphorical wings of a butterfly, but the actual wings of a bat that sparked a chain of events eventually leading to the global COVID-19 pandemic. The crisis has swept across the planet, demonstrating the global interconnectedness of business, pleasure, and politics [38]. Numerous leadership studies include comments on leaders' personalities and especially the need to be positive, e.g., extraversion with positive energy, being inspirational, expressing confidence, being charismatic [39, 40]. And this desire for positivity seems to be universal. Ideal leaders everywhere in the world are expected to develop a vision, inspire the masses, and develop a successful performance-oriented team [41]. Books like "How to Be a Positive Leader" [42] advise us that being optimistic is important, even during tough times. Maybe especially during a crisis, leaders need to encourage and persuade their followers, and in doing so what they essentially require is to exude positivity and be confidence builders. Positive leaders achieve better results. Business leaders who curtail their activities and dismiss employees are expected to do so while putting on a hopeful face and presenting a favorable scene of the future. This was the advice in previous economic downturns, and this will again be the guidance now. But do inspirational leaders, those who stay positive despite a global pandemic, provide the best version of leadership to get us out of this grave mess and toward the best possible outcomes? Researchers across the globe have explored a wide range of crises and the environments in which they befall, in addition to the inherent risks extant in organizational and leadership, practices from a variety of perspectives, focusing on man-made threats [43–45], but there are little data on leadership challenges and organizational consequences ascending from various kinds of threats like COVID-19 [46–48]. Employee productivity is down during this era of instability, workers are partly affected by unemployment, and the staff goes through tough times when they are unprepared to respond to the current circumstance. As part of crisis management, frequent, and clear communication between the leader and staff has a positive effect on the organization's success, providing perceived security at work as well as in the personal lives of their employees. The leader must possess constructive accountability, strong environmental and situational understanding, and the ability to combat disinformation efficiently with sound arguments. To maximize loyalty and devotion to the organization, the leader must first understand the qualities of his subordinates, admire them, and affirm them. Simultaneously, the leader must be agile, prioritize needs other than respond impulsively, and communicate knowledge effectively on all levels by swift adaptability.

Since the innovation process has several distinct features, innovation leadership differs from leadership in other organizational contexts [49]. The COVID-19 pandemic has turned into an innovation catalyst, sparking a surge of innovative problem-solving in all industries. To succeed, spirited businesses have promptly adapted to changing market demands and developments, as well as engaged in continuous learning. While it may be alluring to slow down innovation efforts during a pandemic, but it is the right opportunity to double down so. It is the most appropriate time to make a decision. In the face of increased uncertainty, leaders are more likely to concentrate on operations to reduce costs and increase profit margins. However, it is particularly in these cases that innovation can be prioritized. The rising leaders in today's crisis will facilitate invent solutions to tomorrow's problems. Leadership is much more important in times of confusion and chaos because the leader decides about how an organization reacts. Businesses must be led by a steady and reassuring side, but they must still pivot and be flexible as the situation calls for it.

Unlocking the Mask: A Look at the Role of Leadership 93

Grit, perseverance, and collaboration are some of the characteristics of such a culture that underscores innovation, and these are required to meet some of the most pressing issues. Whilst companies and other organizations all around the world confront crises, a few ways that leaders can help to promote an innovative culture are given here [50].

Learnability is a must: In these times of massive transition, learning rapidly is more critical than ever. Leaders must have a high learnability quotient, besides high IQ (intelligence quotient) and EQ. The challenges of today are so complicated and multi-faceted that they necessitate learning as situations change. Leaders must set an example by searching out new opportunities, and unusual viewpoints, and showing openness to newer ideas while fostering a learning culture in their organizations.

Experimentation and Communication: Leaders must understand the technological expertise needed to successfully adapt their company in times of transition, such as how to migrate to remote work, while we are in uncharted terrain. They should be surrounded by experts and devote time to staying up to date on the new developments, threats, and opportunities confronting their business, as well as how to convey those needs effectively throughout the employees.

Ability to adjust for the Subsequent Crisis: Neither Crisis nor transformation are one-time events. Since change is inevitable, it is critical to be flexible and adaptable. Feedback is critical in determining what actions leaders should take. Leaders must be quick to innovate, explore, and learn, and they must encourage their teams to do so as well. The world of tomorrow is just getting started.

5.5 DISCUSSION

Despite the government's efforts, much more could have been accomplished much earlier if short, medium, and long-term strategic plans had been implemented. COVID-19 has taught every country, including India, the value of successful healthcare planning and preparations in the event of a potential pandemic. Distancing yourself from society is a deterrent, but lockdowns can be harmful to the economy. To ensure that emergency health services can deal with the pandemic more efficiently, better cooperation between government officials and the local healthcare workforce is needed [1]. The pandemic had ravaged the world and led to a huge number of deaths irrespective of national borders. While the percent of deaths was higher in less developed economies, the shared experience of helplessness changed how people in the industrialized world viewed their poorer brethren. The coronavirus could have killed them too, easily. This realization led to empathy toward the other that heretofore had been lacking when it came to relations between the haves and have-nots. For the first time, We Are the World became meaningful beyond being an idealistic song title. In an organizational crisis, leadership is important, and it is often theorized as the technique of exercising social influence [51]. Leaders must not only mount an appropriate tactical response to an organizational crisis but also meet a symbolic need for direction and advice from its members [36]. It is tempting to dream of putting things "back to normal," but it appears to be quite far from reality. As a matter of fact, millions of people do not aspire to or deserve to return to "normal." The contemporary pattern has writhed to bring in health, wellbeing, and prosperity to many people. Now that the "old" system's lack of stability, as well as our potential to mobilize large amounts of capital and wealth while the economy is in jeopardy, hopes will be boosted on what more is now possible in the face of future crises.

5.6 CONCLUSION

Although globalization aided in the dissemination of the virus, co-creation with those around the world will aid in the resolution of the crisis. We must use this opportunity to focus on the need to reform and transform our world on a global scale; examine historical lessons and reset our aspirations for the future. The systemic shocks that we are now witnessing and will be experiencing in the future pose many concerns about what we have taken for granted, and show what is achievable when we need to act quickly. With the number of systems shocks we anticipate, now is a critical stage to mull over several key issues. This crisis will harden certain people's firmly held beliefs, whilst it will shape new possibilities and perceptions for others. The truth is that our whole way of life is all set to be perpetually transformed. This is an opportunity to shape the future rather than simply respond to it. This pandemic has clearly delineated that humanity is above all and that leadership is a shared responsibility. Appropriate time to think about what it means to be a leader, regardless of title or rank has arrived. Even though isolation seems like it goes against our biology, we should reach out (from a physical distance) to embrace one another the best we

Unlocking the Mask: A Look at the Role of Leadership

can. We should think carefully about where we get our information and how we respond to it. We should all lead by example, by communicating clearly with one another, and by having a common objective and action plan. Nobody has witnessed such a more critical moment to be kind to ourselves and one another than now. We are, by necessity, in an age of adaptation and evolution. Society, work, and education may be reformed for the better as we develop on the other side.

KEYWORDS

- **COVID-19 pandemic**
- **creativity**
- **innovation**
- **innovation in crisis**
- **institute for the future**
- **leadership amid pandemic crisis**
- **leadership in crisis**
- **Middle East respiratory syndrome**

REFERENCES

1. Haque, A., (2020). The COVID-19 pandemic and the public health challenges in Bangladesh: A commentary. *Journal of Health Research, 34*(6), 563–567. doi: 10.1108/jhr-07-2020-0279.
2. Leadership, (2021). *Complexity and Change: Learning from the COVID-19 Pandemic.* UWE Bristol Research Repository Home. https://uwerepository.worktribe.com/output/5973281 (accessed on 5 July 2022).
3. (2021). *Coronavirus Disease (COVID-19)–World Health Organization.* WHO | World Health Organization. https://www.who.int/emergencies/diseases/novel-coronavirus-2019 (accessed on 5 July 2022).
4. *Countries Where Coronavirus Has Spread.* Worldometer - Real-Time World Statistics. Last modified January/February 0051. https://www.worldometers.info/coronavirus/countries-where-coronavirus-has-spread/ (accessed on 5 July 2022).
5. Mbogo, R. W., (2020). Leadership roles in managing education in crises: The case of Kenya during COVID-19 pandemic. *European Journal of Education Studies, 7*(9). doi: 10.46827/ejes.v7i9.3250.

6. Zamoum, K., & Tevhide, S. G., (2018). Crisis management: A historical and conceptual approach for a better understanding of today's crises. *Crisis Management -Theory and Practice.* doi: 10.5772/intechopen.76198.
7. Nohria, N., (2020). What organizations need to survive a pandemic. *Harvard Business Review.*
8. Brown, G., (2020). *In the Coronavirus Crisis, Our Leaders are Failing Us.* The Guardian. https://opensourceventilator.ie/about (accessed on 5 July 2022).
9. Kellerman, B., (2004). *Bad Leadership: What It Is, How It Happens, Why It Matters.* Brighton: Harvard Business Press.
10. Jennifer, S., (2020). *This Is What Happens When a Narcissist Runs a Crisis.* The New York Times (New York). https://www.nytimes.com/2020/04/05/opinion/trump-coronavirus.html (accessed on 5 July 2022).
11. Rachel, S., (2020). *People Are Dying and All Britain Can Talk About Is Boris Johnson.* The New York Time (New York). https://www.nytimes.com/2020/04/16/opinion/coronavirus-boris-johnson.html (accessed on 5 July 2022).
12. American Psychiatric Association, (2013). *Diagnostic and Statistical Manual of Mental Disorders (DSM-5) (5th edn.).* American Psychiatric Association. https://www.appi.org/Diagnostic_and_Statistical_Manual_of_Mental_Disorders_DSM-5_Fifth_Edition (accessed on 5 July 2022).
13. ET Bureau. (2020). *PM Modi Urges Countrymen to Dispel the Darkness Spread by Coronavirus by Lighting a Candle on April 5.* The Economic Times. https://economictimes.indiatimes.com/news/politics-and-nation/pm-modi-urges-countrymen-to-dispel-the-darkness-spread-by-coronavirus-by-lighting-a-candle-on-april-5/articleshow/74959545.cms?from=mdr (accessed on 5 July 2022).
14. Census Provisional Population Totals, (2011). *Data for All.* https://www.dataforall.org/dashboard/censusinfoindia_pca/ (accessed on 5 July 2022).
15. Mendenhall, M. E., Milda, Ž., Günter, K. S., & Clapp-Smith, R., (2020). *Responsible Global Leadership: Dilemmas, Paradoxes, and Opportunities.* London: Routledge.
16. India's COVID-19 emergency, (2021). *The Lancet, 397.* www.thelancet.com (accessed on 5 July 2022).
17. Fleming, S., (2021). *How Big Business is Joining the Fight Against COVID-19.* World Economic Forum. https://www.weforum.org/agenda/2020/03/big-business-joining-fight-against-coronavirus (accessed on 5 July 2022).
18. Maak, T., & Nicola, M. P., (2009). Business leaders as citizens of the world. Advancing humanism on a global scale. *Journal of Business Ethics, 88*(3), 537–550. doi: 10.1007/s10551-009-0122-0.
19. Beechler, S., & Javidan, M., (2007). Leading with a global mindset. *Advances in International Management, 131*–169. doi: 10.1016/s1571-5027(07)19006-9.
20. Paine, L. S., (2006). A compass for decision making. In: Thomas, M., & Nicola, P., (eds.), *Responsible Leadership* (pp. 54–67). London: Routledge.
21. Pless, N. M., (2007). Understanding responsible leadership: Role identity and motivational drivers. *Journal of Business Ethics, 74*(4), 437–456. doi: 10.1007/s10551-007-9518-x.
22. Anne, S. T., (2020). *COVID-19 Crisis: A Call for Responsible Leadership Research.* RRBM Network (blog). https://www.rrbm.network/covid-19-crisis-a-call-for-responsible-leadership-researchanne-s-tsui/ (accessed on 5 July 2022).

Unlocking the Mask: A Look at the Role of Leadership 97

23. Jordan, B. A., Laura, F., Felicitas, J., & Erik, R., (2020). *Innovation in a Crisis: Why It is More Critical Than Ever.* McKinsey & Company. https://www.mckinsey.com/business-functions/strategy-and-corporate-finance/our-insights/innovation-in-a-crisis-why-it-is-more-critical-than-ever (accessed on 5 July 2022).

24. Kozioł-Nadolna, K., (2020). The role of a leader in stimulating innovation in an organization. *Administrative Sciences, 10*(3), 59. doi: 10.3390/admsci10030059.

25. Gumusluoglu, L., & Arzu, I., (2009). Transformational leadership, creativity, and organizational innovation. *Journal of Business Research, 62*(4), 461–473. doi: 10.1016/j.jbusres.2007.07.032.

26. Gates, B., (2018). Innovation for pandemics. *New England Journal of Medicine, 378*(22), 2057–2060. doi: 10.1056/nejmp1806283.

27. Bartik, A. W., Bertrand, M., Cullen, Z., Glaeser, E. L., Luca, M., & Stanton, C., (2020). The impact of COVID-19 on small business outcomes and expectations. *Proc. Natl. Acad. Sci. U S A., 117*(30), 17656–17666. doi: 10.1073/pnas.2006991117.

28. Helfat, C. E., (1997). Know-how and asset complementarity and dynamic capability accumulation: The case of r&d. *Strategic Management Journal, 18*(5), 339–360. doi: https://onlinelibrary.wiley.com/doi/10.1002/%28SICI%291097-0266%28199705%2918%3A5%3C339%3A%3AAID-SMJ883%3E3.0.CO%3B2-7.

29. Martínez-Sánchez, A., María-José Vela-Jiménez, Pérez-Pérez, M., & De-Luis-Carnicer, P., (2011). The dynamics of labor flexibility: Relationships between employment type and innovativeness. *Journal of Management Studies, 48*(4), 715–736. doi: 10.1111/j.1467-6486.2010.00935.x.

30. Marullo, C., Piccaluga, A., & Cesaroni, F., (2020). How to invest in r&d during a crisis? Exploring the differences between fast-growing and slow-growing *SMEs. Piccola Impresa/Small Business, 26*(1). doi: https://journals.uniurb.it/index.php/piccola/article/view/2792.

31. Jover, A. J., Javier, F. M., & Víctor, J. M., (2005). Flexibility, fit and innovative capacity: An empirical examination. *International Journal of Technology Management, 30*(1, 2), 131. doi: 10.1504/ijtm.2005.006348.

32. Aghion, P., Stefan, B., Lea, C., & Holger, H., (2018). The causal effects of competition on innovation: Experimental evidence. *The Journal of Law, Economics, and Organization, 34*(2), 162–195. doi: 10.1093/jleo/ewy004.

33. Blake, D., & William, B., (2001). Survivor bonds: Helping to hedge mortality risk. *The Journal of Risk and Insurance, 68*(2), 339. doi: 10.2307/2678106.

34. Zhang, X., & Zhou, K., (2019). Close relationship with the supervisor may impede employee creativity by suppressing vertical task conflict. *R&D Management, 49*(5), 789–802. doi: 10.1111/radm.12375.

35. Elizabeth, A., (2020). *Why Do Innovation and Creativity Thrive in Crisis?* Babson Thought & Action. https://entrepreneurship.babson.edu/innovation-and-creativity-in-a-crisis/ (accessed on 5 July 2022).

36. Boin, A., & Paul, H., (2003). Public leadership in times of crisis: Mission impossible? *Public Administration Review, 63*(5), 544–553. doi: 10.1111/1540-6210.00318.

37. Mumford, M. D., Ginamarie, M. S., Blaine, G., & Jill, M. S., (2002). Leading creative people: Orchestrating expertise and relationships. *The Leadership Quarterly, 13*(6), 705–750. doi: 10.1016/s1048-9843(02)00158-3.

38. Osland, J. S., Mendenhall, M. E., Reiche, B. S., Szkudlarek, B., Bolden, R., Courtice, P., Vaiman, V., et al., (2020). Perspectives on global leadership and the COVID-19 crisis. In: *Advances in Global Leadership* (pp. 3–56). Emerald Publishing Limited. https://www.emerald.com/insight/content/doi/10.1108/S1535-120320200000013001/full/html.

39. Burns, J. M., (2010). *Leadership* [originally published 1978]. New York, NY: Harper Perennial Political Classics.

40. Judge, T. A., Joyce, E. B., Remus, I., & Megan, W. G., (2002). Personality and leadership: A qualitative and quantitative review. *Journal of Applied Psychology, 87*(4), 765–780. doi: 10.1037/0021-9010.87.4.765.

41. Dorfman, P., Mansour, J., Paul, H., Dastmalchian, A., & Robert, H., (2012). GLOBE: A twenty year journey into the intriguing world of culture and leadership. *Journal of World Business, 47*(4), 504–518. doi: 10.1016/j.jwb.2012.01.004.

42. Geiger, D., Jane, E. D., & Gretchen, M. S., (2016). How to be a positive leader: Small actions, big impact. San Francisco, CA: Berrett-Koehler Publishers, Inc., $26.95 softcover. *Personnel Psychology, 69*(3), 213, 763–766. doi: 10.1111/peps.12174.

43. Boin, R. A., (2005). Crisis to disaster: Toward an integrative perspective. In: Perry, R. W., & Enrico, L. Q., (eds.), *What is a Disaster?: New Answers to Old Questions.* Bloomington: Xlibris Corporation.

44. Kayes, C., Allen, N., & Self, N., (2017). How leaders learn from experience in extreme situations: The case of the US military in Takur Ghar, Afghanistan. In: *Leadership in Extreme Situations* (pp. 277–294). Springer, Cham.

45. Adkins, C. L., Werbel, J. D., & Farh, J. L., (2001). A field study of job insecurity during a financial crisis. *Group & Organization Management, 26*(4), 463–483.

46. Jack, M., (2020). *4 COVID-19 Leadership Lessons.* Chief Executive.net. https://chiefexecutive.net/4-covid-19-leadership-lessons/ (accessed on 5 July 2022).

47. Tracy, B., (2020). *5 Predictions About How Coronavirus Will Change The Future of Work.* Forbes. https://www.forbes.com/sites/tracybrower/2020/04/06/how-the-post-covid-future-will-be-different-5-positive-predictions-about-the-future-of-work-to-help-your-mood-and-your-sanity/?sh=1d5575e63e22 (accessed on 5 July 2022).

48. Halper-Bogusky, K., (2020). *STUDY: Organizations Rising to the Challenge of COVID-19 Communications, but Needs Persist; Leaders Must Address Concerns and Demonstrate Transparency, Clarity and Openness.* https://www.businesswire.com/news/home/20200403005278/en/STUDY-Organizations-Rising-to-the-Challenge-of-COVID-19-Communications-but-Needs-Persist-Leaders-Must-Address-Concerns-and-Demonstrate-Transparency-Clarity-and-Openness (accessed on 5 July 2022).

49. Mumford, M. D., (2000). Managing creative people: Strategies and tactics for innovation. *Human Resource Management Review, 10*(3), 313–351. doi: 10.1016/s1053-4822(99)00043-1.

50. Manpower group, (2021). *How Leaders Can Foster Innovation in Times of Crisis.* ManpowerGroup-Workforce-Resources (blog). https://workforce-resources.manpowergroup.com/blog/how-leaders-can-foster-innovation-in-times-of-crisis (accessed on 5 July 2022).

51. Mumford, M. D., Tamara, L. F., Jay, J. C., & Cristina, L. B., (2007). Leader cognition in real-world settings: How do leaders think about crises? *The Leadership Quarterly, 18*(6), 515–543. doi: 10.1016/j.leaqua.2007.09.002.

CHAPTER 6

Emotional Intelligence as a Tool to Manage Conflict, Emotions, and Behavior of Human Beings During the Pandemic COVID-19

SHRUTI TRAYMBAK,[1] MEGHNA SHARMA,[2] SHUBHAM AGGARWAL,[1] and KRITY GULATI[3]

[1]*Department of Management, Lloyd Business School, Greater Noida, Uttar Pradesh, India, E-mail: shruti@lloydcollege.in (S. Traymbak)*

[2]*AIBS, Amity University, Noida, Uttar Pradesh, India*

[3]*Department of Management, LIMT, Greater Noida, Uttar Pradesh, India*

ABSTRACT

The objective of the present study is to examine the effect of factors of Emotional Intelligence like Regulations of Emotions (ROE), Self-Emotion Appraisal (SEA), Other's Emotional Assessment (OEA), Use of Emotion (UOE) and Regulation of Emotion (ROE) on Satisfaction with Life (LS). Apart from this, the present study also examines the moderating effects of Gender between Emotional Intelligence and Satisfaction with Life. The present study used linear regression, stepwise regression, and interactive effects, ROE and UOE have significant positive impact on LS. UOE which refers to use emotions and feelings in constructive way and Regulation of Emotions (ROE) which helps an individual to cope up with psychological stress and satisfied with life as well. Interestingly gender has no moderating effects.

Emotional Intelligence for Leadership Effectiveness: Management Opportunities and Challenges During Times of Crisis. Mubashir Majid Baba, Chitra Krishnan, & Fatma Nasser Al-Harthy (Eds.)
© 2023 Apple Academic Press, Inc. Co-published with CRC Press (Taylor & Francis)

6.1 INTRODUCTION

The term "emotional intelligence (EI)" came into the picture in 1964, coined by Michael Beldoch. Emotional Intelligence (EQ/EI) is one of the prominent topics and has received great attention over the last two decades [14].

Emotional intelligence or EQ/EI elicits abilities to connect intelligence, empathy, and emotions to understand own's and other's feelings. There are various models that define emotional intelligence (EI/EQ) such as Goleman's [8] EI performance model, Bar-On's [2] EI competencies and Mayer et al. [6] defined EI as an ability model which are based on cognitive skills, non-cognitive skills, and competencies that help us to understand own's emotions and other's emotions in a positive manner and guide human behavior. According to Goleman [8] emotional intelligence (EQ/EI) consist of five factors like self-awareness refers to understand other's and own feelings, self-regulations mean to keep one's own emotions calm in difficult situations and to maintain calmness irrespective of one's emotions, motivation refers to operate with hope of success, empathy means able to understand the emotions and feelings of others and provide affective or emotional support when required and social skills is related to handle the problems and emotional conflict with diplomacy. Out of which two of these factors are linked to personal competencies, such as recognizing and regulations while rest two are related to social competencies, such as social awareness and relationship management. Both competencies personal and social competencies help in resolving conflict at home during COVID-19 by recognizing how your behavior impacts others, awareness of himself or herself or other's emotions, caring about others what others are going through, hearing what other person is saying and handling conflict successfully.

According to Bar-On's [2] EI ability model believe in guiding human behavior, stress management, and maintain good interpersonal relationship. Mayer et al.'s [6] ability model of EI emphasized on right understanding, how to control emotions, enhanced thinking, and effort to bring more intelligent method in order to strengthen the relationship. Salovey and Mayer [16] found that individuals who have high EI, he or she has the power to analyze own's and other's emotions and regulate emotions as well.

During pandemic COVID-19 work from home, lay-off, and salary deduction changed the human behavior which led to change in psychological imbalance, people suffering from depression, anxiety, and stress at home. In such pandemic situation life satisfaction (LS) is very less. EI actually affects and influences all aspects of our lives such as managing conflicts. EI/EQ focuses on trust, respect, excellence, growth, happiness, and harmony that result into right understanding and good relationship. Understanding the distinctive attributes of emotions and EI is extremely important. Emotion is a natural and instinctive mental state which stems from our present and past experiences. On the other side, EI has potential to resolve conflicts in every aspect of life.

In this study, the association between EI and LS was examined with the use of self-reports. [3, 4]. A four-factor scale based on emotional intelligence (EQ/EI) was employed in the present study of Davies et al. [1] and the concept of Salovey and Mayer [16]. This scale evaluates four dimensions of EI: self-emotion appraisal (SEA), other's emotional assessment (OEA), use of emotion (UOE), and management of emotion (ROE).

The term SEA refers to a person who has the ability to comprehend and examine their emotions, as well as the ability to show them naturally. Similarly, other's emotional appraisal (OEA) means to understand other's feelings and emotions. UOE acknowledges an individual has a potential to use emotions in creative and constructive work and regulation of emotion (ROE) helps an individual to revive from psychological stress. All four factors are very important in resolving conflict and in LS.

A very few researchers explored the relationship of four dimensions of WLEIS with LS and the interactive effect of emotional intelligence (EI/EQ) on LS in the Indian context to manage conflict and stress. The objective of present research is to identity that which dimensions of WLEIS is more effective and efficient to understand other's emotions, resolve conflicts and problems. Apart from this which factor is more significantly correlated with LS.

6.2 METHODOLOGY

6.2.1 SAMPLE

The present study considered a sample of 93 as shown in Table 6.1 showing respondents characteristics such as gender, education, and experience.

TABLE 6.1 Details of Respondents

Characteristics of Respondents	Number	Percentage (%)
Gender		
Male	50	53
Female	43	46
Education		
Doctorate	3	3
Undergraduate	4	4
Graduate	15	16
Postgraduate	71	76
Experience		
0–5 years	65	69
5–10 years	11	11
10–15 years	4	4
15–20 years	4	4
More than 20 years	9	22

6.3 MEASUREMENT SCALE

The study used self-report WLEIS scale (2002) to measure EI of four factors. This scale consists of 16 items and four items per factor [5]. Satisfaction with life scale (SWLS) that consists of five items to measure LS. An item of LS is "I am satisfied with my life." The items were measured on a 5-point Likert scale.

6.4 RESULTS AND DISCUSSION

6.4.1 RELIABILITY AND VALIDITY

The present used SPSS statistics to verify the reliability of all 21 items which is more than 0.7, i.e., Cronbach's Alpha is 0.895 as shown in Table 6.2. According to Nunally [13] for reliability Cronbach alpha should be 0.7 or more than 0.7.

Emotional Intelligence as a Tool to Manage Conflict 103

TABLE 6.2 Reliability

Cronbach's Alpha	Cronbach's Alpha	Items
0.895	0.900	21

6.4.2 LINEAR REGRESSION

The Linear Regression showed that SEA, OEA, UOE, and ROE has impact of 0.263 on LS (Tables 6.3 and 6.4).

TABLE 6.3 Linear Regression Output

Model Summary				
Model	R	R^2	Adj R^2	Std. Error
1	0.513[a]	0.263	0.230	0.70317

TABLE 6.4 Impact of Four Independent Factors on Dependent Factor Life Satisfaction

Model		UnStd. Coefficients			t	Sig.
		B	Std. Error	β,		
1		0.362	0.556		0.651	0.517
	SEA	0.033	0.143	0.028	0.231	0.818
	OEA	0.200	0.131	0.165	1.527	0.130
	UOE	0.256	0.147	0.216	1.741	0.085
	ROE	0.232	0.094	0.261	2.454	0.016

[a]Dependent variable: LS, whereas independent variable: SEA, OEA, UOE, and ROE.

6.4.3 STEPWISE REGRESSION

The present study found that UOE and ROE are the important factors to predict LS. They have a significant impact on LS as shown in Tables 6.5 and 6.6.

TABLE 6.5 Stepwise Regression

Model	Taken Variables	Not Taken Variables	Method
1	UOE	.	Stepwise regression
2	ROE	.	

[a]Dependent variable: LS.

TABLE 6.6 Impact of Two Important Factors

Model Summary									
Model	R	R²	Adjusted R²	Std. Error of the Estimate	Change Statistics				
					R² Change	F Change	df1	df2	Sig. F Change
1	0.412[a]	0.170	0.161	0.73402	0.170	18.634	1	91	0.000
2	0.493[b]	0.243	0.226	0.70472	0.073	8.724	1	90	0.004

[a]Predictors: (Constant), UOE.
[b]Predictors: (Constant), UOE, ROE.

6.4.4 INTERACTIVE EFFECT OF EMOTIONAL INTELLIGENCE (EQ/EI)

The present study used the interactive effect of EI with all four factors and found insignificant impact to predict LS. Interestingly it has been found that all in linear regression, stepwise regression, and interactive effect ROE and UOE has significant positive impact on LS. Gender has no significant impact between SEA, OEA, ROE, UOE, and LS as shown in Tables 6.7 and 6.8.

TABLE 6.7 Interactive Effects of EI_OEA_SEA_UOE_ROE

Model		UnStd Coefficients		Std Coefficients	t	Sig.
		B	Std. Error	β		
1	(Constant)	0.362	0.556	–	0.651	0.517
	SEA	0.033	0.143	0.028	0.231	0.818
	OEA	0.200	0.131	0.165	1.527	0.130
	UOE	0.256	0.147	0.216	1.741	0.085
	ROE	0.232	0.094	0.261	2.454	0.016
2	(Constant)	–0.278	0.910	–	–0.306	0.761
	SEA	0.077	0.151	0.066	0.509	0.612
	OEA	0.248	0.142	0.204	1.749	0.084
	UOE	0.307	0.158	0.260	1.943	0.055
	ROE	0.311	0.130	0.349	2.395	0.019
	EI_OEA_SEA_UOE_ROE	–0.001	0.001	–0.179	–0.889	0.376

[a]Dependent variable: LS.

Emotional Intelligence as a Tool to Manage Conflict 105

TABLE 6.8 Interactive Effects of Gender

Model Summary[d]									
Model	R	R^2	Adjusted R^2	Std. Error	Change Statistics				
					R^2 Change	F Change	df1	df2	Sig. F Change
1	0.513[a]	0.263	0.230	0.70317	0.263	7.866	4	88	0.000
2	0.515[b]	0.265	0.223	0.70632	0.002	0.218	1	87	0.642
3	0.520[c]	0.271	0.220	0.70781	0.005	0.635	1	86	0.428

[a]Predictors: (Constant), ROE, OEA, SEA, UOE.
[b]Predictors: (Constant), ROE, OEA, SEA, UOE, Gender.
[c]Predictors: (Constant), ROE, OEA, SEA, UOE, Gender, EI_Gender.
[d]Dependent Variable: LS.

6.5 CONCLUSION

On the basis of the above empirical analysis like linear regression, stepwise regression, and interactive effects, ROE and UOE have significant positive impact on LS. UOE which refers to use emotions and feelings in constructive way and Regulation of Emotions (ROE) which helps an individual to cope up with psychological stress and satisfied with life as well. Interestingly gender has no moderating effects between SEA, ROE, UOE, OEA, and LS. On the basis of empirical findings, ROE and UOE play an important to manage conflicts among males and females.

KEYWORDS

- emotional intelligence
- life satisfaction
- management of emotion
- other's emotion appraisal
- regulation of emotion
- self-emotion appraisal
- use of emotion

REFERENCES

1. Davies, M., Stankov, L., & Roberts, R. D., (1998). Emotional intelligence: In search of an elusive construct. *Journal of Personality and Social Psychology, 75,* 989–1015.
2. Bar-On, R., (1997). *The Emotional Quotient Inventory (EQ-i): A Test of Emotional Intelligence.* Toronto: Multi-Health Systems.
3. Charbonneau, D., & Nicol, A. A., (2002). Emotional intelligence and leadership in adolescents. *Personality and Individual Differences, 33,* 1101–1113.
4. Ciarrochi, J. V., Deane, F., & Anderson, S., (2002). Emotional intelligence moderates the relationship between stress and mental health. *Personality and Individual Differences, 32,* 197–209.
5. Diener, Ed., Emmons, R. A., Larsen, R. J., & Griffin, S., (1985). The satisfaction with life scale. *Journal of Personality Assessment, 49,* 71–75.
6. Mayer, J. D., Caruso, D., & Salovey, P., (1999). Emotional intelligence meets traditional standards for an intelligence. *Intelligence, 27,* 267–298.
7. Matthews, G., Zeidner, M., & Roberts, R. D., (2012). Emotional intelligence: A promise unfulfilled? *Japanese Psychological Research, 54*(2), 105–127.
8. Goleman, D., (1995). *Emotional Intelligence.* New York: Bantam Books.
9. Goleman, D., (1998). What makes a leader? *Harvard Business Review.*
10. Goleman, D., (2000). *Working with Emotional Intelligence.* New York: Bantam Books.
11. Salovey, P., Marc, A. B., & John, D. M., (2004). *Emotional Intelligence.* Port Chester, N.Y.: Dude Pub.
12. Mayer, J. D., Roberts, R. D., & Barasade, S. G., (2008). Human abilities: Emotional intelligence. *The Annual Review of Psychology, 59,* 507–536.
13. Nunnally, J. C., (1978). *Psychometric Theory.* New York: McGraw-Hill.
14. Matthews, G., Zeidner, M., & Roberts, R., (2002). *Emotional Intelligence: Science and Myth?* Cambridge, MA: MIT Press.
15. Salovey, P., Mayer, J., & Caruso, D., (2004). Emotional intelligence: Theory, findings, and implications. *Psychological Inquiry,* 197–215.
16. Salovey, P., & Mayer, J. D., (1990). Emotional intelligence. *Imagination, Cognition, and Personality, 9,* 185–211.
17. Salovey, P., & Mayer, J. D., (1989). Emotional intelligence. *Imagination, Cognition, and Personality, 9*(3), 185–211.
18. Wong, C. S., & Law, K. S., (2002). The effects of leader and follower emotional intelligence on performance and attitude: An exploratory study. *The Leadership Quarterly, 13,* 243–274.

CHAPTER 7

Thriving in the "New Normal": An Era of COVID-19

MOHD. ZIA UL HAQ RAFAQI[1] and ZAINAB MUSHEER[2]

[1]*Department of Education University of Kashmir, South Campus, Anantnag, Jammu and Kashmir, India, E-mail: mzrafiqi@gmail.com*

[2]*Department of Education, Aligarh Muslim University, Aligarh, Jammu and Kashmir, India*

ABSTRACT

The study investigates the threatened autonomy on the psychological retrieval process of college employees who were called to work soon after unlock process started in India when the ongoing COVID-19 stressor was still there. This study focuses on the need for the development of autonomy to elicit goal-oriented behaviors so that normalcy could be back in behavior when it is totally distorted during the time of the COVID-19 pandemic. Helplessness and reduced authenticity are the two indicators of weak autonomy that have been used to examine the process of retrieval soon after unlock-I and how it changed after unlock-VI. The data were collected in two phases through an experience sampling dataset. The first phase is from a period of two-weeks 1 June to 15 June 2020 and the second phase 19 Nov 2020 to 30 Nov 2020. The study will highlight the path for the psychological retrieval process in case of helplessness and reduced authenticity among employees with neuroticism personality characteristics. This study provides novel ideas for the psychological well-being of the

---

Emotional Intelligence for Leadership Effectiveness: Management Opportunities and Challenges During Times of Crisis. Mubashir Majid Baba, Chitra Krishnan, & Fatma Nasser Al-Harthy (Eds.)
© 2023 Apple Academic Press, Inc. Co-published with CRC Press (Taylor & Francis)

108   *Emotional Intelligence for Leadership Effectiveness*

employees and suggests ways for beginning the psychological retrieval process during the COVID-19 pandemic.

7.1 INTRODUCTION

The sudden outbreak of novel coronavirus (nCoV) "SARS-CoV-2" across the globe has been declared as a state of "global emergency" leading to closure of public and private institutions intended to restrict the spread of the virus [1]. This virulent disease devours a massive impact on public health, leading to a huge economic crisis all over the world. After a sharp global upsurge in the figure of the COVID-19 virus, several countries endured strict stratagem to curtail its spread, a "nationwide complete lockdown" being one such measure [2]. Likewise, on 24 March 2020, the government of India had declared a "nationwide complete lockdown," which was initially supposed to be of 21 days and was later on increased further [3]. People were forced to "remain at home" and "all the offices, shopping malls and educational institutions" were completely shut down for the initial 21 days. Although this "lockdown" transpired to control the spread of coronavirus but it devised a huge emotional ramification upon the minds of both young and old generation [4–7]. Studies have shown an increase in the level of stress, anxiety, depression plus the poor state of living during this time [8–11]. Despite the steep increase in COVID cases in India there appeared a silver lining in the dark cloud when the government started the unlock process. The process of unlock—I had started in India from 1 June 2020 in the face of rampaging pandemic [12]. Although, the educational institutions remained closed but the staff was called to carry out the office work initially for a few hours and on odd-even basis. College employees living in the containment zones were strictly prohibited from entering the campus. This unlock-I process was strenuous challenging task, as critical risk trade-off between health and economics. This unlock was followed by many other unlocks finally leading to unlock-VI in India. Unlock six was announced from 1 November 2020. Up till this time the COVID-19 stressor was still there, and people had become restless sitting at their homes. Fear of transmission of diseases and losing of jobs were two extreme phobias in the mind of the working class. Researches had confirmed that COVID-19 is considered as the most stressful period in the entire career of employees globally [13]. Such a situation not only damages the general wellbeing of the employees but also damages the sense of autonomy in them [14, 15]. Autonomy is the state where an individual work on self-belief and values without any external control. Helplessness and reduced authenticity are the two indicators of weak autonomy that are used to examine the process of retrieval soon after unlock one and how it changed after unlock six. Studies from the past have shown that the psychological retrieval process differs from individual to individual and there way of coping with the stressor [13, 17]. These studies were mostly conducted on the psychological retrieval after the stressor had completely extinct [18, 19]. The present study provides valuable insight on how the psychological retrieval process starts even when the stressor still exists. Research in the past have shown that autonomy is essential for the psychological wellbeing even before the arrival of coronavirus (CoV) [20]. The study highlights the path for the psychological retrieval process in case of helplessness and reduced authenticity among employee of neuroticism personality characteristic. People with high neuroticism are at the risk of developing more anxiety and mood disorders compared with those who are low on neuroticism index [21]. A perusal of literature highlights the significance of strong autonomy in the development of psychological retrieval process [22], but not much work has been done to examine employees respond to time when antagonized with "weak autonomy." Most importantly, when a man is "deprived of autonomy" it gets increasingly accessible for the need of behavioral schema to look into [23, 24]. During the COVID-19 pandemic workers may retort to helplessness by enthralling in virtual conferencing through software like WebEx, Zoom, Google meet, etc., and by creating new work place at home. Employees may reduce the sense of autonomy like expressing an individual's personality in a virtual work environment. The majority of work in the past addressed the autonomy restorative response in real life setup in a more controlled environment [25]. The investigators explore the manifestations of threatened autonomy: helplessness and reduced authenticity on the psychological retrieval process of college employees even when the stressor is there. The hypotheses for the present study are built on the "autonomy restoration" theory [26] which postulates the demand to restore an individual's autonomy shows "immediately following the experience of an autonomy-depriving event." It is assumed in the present study that after exposure to ongoing COVID-19 stressor helplessness and reduced authenticity (inaccuracy) decreases over time among employees with low neuroticism characteristics.

7.2 METHODS

7.2.1 RESEARCH DESIGN

A qualitative inquiry was carried out in the present study using structured interview with the employees working in various colleges of India. Structured interview was conducted using telephone, video calls and personally approaching the respondents. The data was collected over two phases through the experience sampling method (ESM). The first one for a period of two weeks from 1st June to 15th June 2020 and the second one from 19th Nov 2020 to 30th Nov 2020. The data was collected in total for a period of 26 days, measuring the change in the psychological retrieval process among college employees during Unlock-I and Unlock-VI.

7.2.2 PARTICIPANTS

The participants were the employees working in various colleges of India. The study included both teaching and non-teaching employees of the colleges. All employees corresponded agreed to participate in the present study. The investigation was conducted on a sample of 46 college employees' (20 males and 26 females belonging to the age group of 25–60) selected using convenience and snowball sampling techniques. Respondents were asked certain structured questions two times a day via telephone, video calls, or by personally approaching them for a period of two weeks from 1st June to 15th June 2020 and the second one from 19th Nov 2020 to 30th Nov 2020.

7.2.3 MEASURE

The first part of the interview covers the demographic information like gender, age, experience of work, marital status, job designation, and daily office timing. The second part involves questions like the idea of social isolation, physical distancing, quarantine, restriction on people movement, SOPs followed at the workplace, incorrect, and misleading information, helplessness, etc. The third part includes questions related to stress, loneliness, emotional chaos, anxiety, depression, obsessive-compulsive neurosis, etc. This structured interview was divided into two sets: one for morning and one for evening on a daily basis.

Thriving in the "New Normal": An Era of COVID-19 111

7.2.4 DATA ANALYSIS

The structure interviews were analyzed with regard to the objective of the study. Most of the questions were pre-coded, and some of them were post coded. In order to analyze the effect of neuroticism (lower *vs.* higher), helplessness, and reduced authenticity on psychological retrieval process during COVID-19 unlock one and unlock six in India random-intercept and multi-level regression analyzes were applied. The multi-level regression model depicts the design of the current study with reference to the measurement position (level 1) build within persons (level 2). Altogether the predictors at the first level were "centered within participants" and the predictor of age at the next level was "grand-mean centered" [27]. Moreover, to measure intra class correlation coefficient (ICC), a base model was run in the statistical software without any predictor. Moreover, fixed vs. random model were measured to check if the random model explains the additional variance in comparison to the fixed models. Although the interaction with reduced authenticity showed with higher neuroticism vs. lower neuroticism providing a significant development associated to the fixed model. Additionally, to avoid the risk of overfitting the researcher involved merely two-way interactions, with the exception of reduced authenticity by lower *vs.* higher neuroticism interaction. Moreover, the researcher had to standardize the weak autonomy (helplessness and reduced authenticity) to dodge extremely high eigenvalues in calculations. The final model is exhibited underneath:

Level 1 (within person): Psychological Retrieval Process$_{ti}$ = π_{0i} + π_{1i} Helplesness$_{ti}$ + π_{2i} reduced authenticity$_{ti}$ + π_{3i} Neuroticism (higher vs. lower)$_{ti}$ + π_{4i} Helplessness*Reduced authenticity$_{ti}$ + π_{5i} Helplessness*Neuroticism$_{ti}$ + e_{ti}

Level 2 (between persons): π_{0i} = β_{00} + β_{01} Age + r_{0i}

Level 2 (between persons): π_{1i} = β_{10} + r_{1i}; π_{2i} = β_{20} + r_{2i}; π_{3i} = β_{30} + r_{3i}; π_{4i} = β_{40} + r_{4i}; π_{5i} = β_{50} + r_{5i}

7.3 RESULTS

7.3.1 CHANGE OVER TIME

For analyzing the trend over time with the day of assessment, random-intercept slopes multilevel models were employed. Psychological retrieval process made progress marginally over the 26-days' time slot (β = 0.05, 95%

CI =0.03, 0.09 p < 0.001), reduced authenticity remained constant (β < 0.01, 95% CI = –0.04, 0.04, p = 0.99) and helplessness lowered (β = –0.05, 95% CI = –0.09, –0.03, p < 0.001) (Figure 7.1).

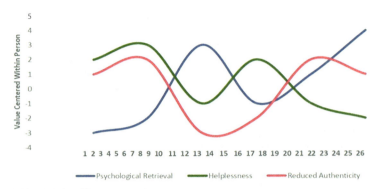

FIGURE 7.1 Change over time.

7.3.2 CROSS-SECTIONAL CORRELATIONS

Finding of the result revealed that there is a negative correlation between psychological retrieval process and helplessness. It also showed psychological retrieval process was negatively correlated with reduced authenticity. Psychological retrieval process was positively associated with lower neuroticism and not higher neuroticism and also helplessness was negatively associated with higher neuroticism. Greater feeling of helplessness was related with more reduced authenticity, and more reduced authenticity was associated with lower neuroticism, which means reduced authenticity predominantly existed among higher neuroticism. Moreover, younger employees showed better psychological retrieval, more authenticity, and lower neuroticism (Table 7.1).

7.3.3 MULTI-LEVEL ANALYZES

ICC value for psychological retrieval process was found to be 53%. This can be explained as 47% variance existed within participants and 53% of variance existed between participants. Further there existed no gender difference in the study with reference to other variables. Thus, gender was not included as a predictor in the study. Gender effects: psychological

Thriving in the "New Normal": An Era of COVID-19   113

retrieval process: $\beta = -2.24$, $p > 0.05$, helplessness: $\beta = -4.21$, $p > 0.05$, reduced authenticity: $\beta = -0.47$ $p > 0.05$. The main effects came to be significant (Table 7.2). Feeling helpless and having more reduced authenticity was associated with poorer psychological retrieval process, but lower neuroticism was associated with improved psychological retrieval process. Lower neuroticism raised the psychological retrieval process by 3.4 units on an average and one unit more of reduced authenticity every day lowered psychological retrieval process by 0.4 units on an average. In terms of interactions, only helplessness among higher vs. lower neuroticism interaction is significant. In a more precise sense, the relation between helplessness and psychological retrieval process was poorer among lower neurotic personality than among higher neuroticism personality.

TABLE 7.1   Cross-Sectional Correlation

Psychological Retrieval Process	1	2	3	4	5
Helplessness	-0.47^{***}				
Reduced authenticity	-0.22^{***}	0.19^{**}			
Higher neuroticism	0.12	-0.18^{**}	0.09		
Lower neuroticism	0.22^{***}	-0.12	-0.38^{***}	0.21^{**}	
Age	-0.18^{*}	0.09	0.33^{***}	0.08	-0.44^{***}

$^{***}p < 0.001$; $^{**}p < 0.01$; $^{*}p < 0.05$.

Weak autonomy (helplessness and reduced authenticity) was standardized. Reference category for "lower neuroticism" was "higher neuroticism"

7.4   DISCUSSION

To examine the impact of hypothesized antecedents on psychological retrieval process during unlock-I and unlock-VI COVID-19 in India an ESM design was used. For explanatory purpose, interactive effects of lower neuroticism, helplessness, reduced authenticity, and age of the college employees on the psychological retrieval process was examined but the effect were mostly null except the interaction between helplessness and lower neuroticism. Based on the above findings, certain implications were highlighted by the researcher.

At the first, the result showed that the employees with lower neuroticism characteristic is related with the improved psychological retrieval

TABLE 7.2   Multi-Level Analyzes for Psychological Retrieval Process During Unlock-I and Unlock-VI of COVID-19 in India

	Random		Fixed			
	Coeff.	SD	Coeff.	B [CI]	SE	t
Intercept within-person	r_{0i}	16.03	β_{00}	68.06 [66.20, 69.91]	0.94	74.0***
Helplessness	r_{1i}	5.34	β_{10}	−5.31[−6.19, −4.59]	0.42	−12.3***
Reduced authenticity	r_{2i}	1.27	β_{20}	−0.41[−0.43, −0.11]	0.21	−3.9**
Higher vs. lower neuroticism	r_{3i}	3.32	β_{30}	3.43 [2.26, 4.39]	0.44	9.5***
Helplessness* reduced authenticity	r_{4i}	0.49	β_{40}	−0.02 [−0.11, 0.18]	0.07	0.33
Helplessness* higher vs. lower neuroticism	r_{5i}	3.52	β_{50}	1.59 [0.59, 2.42]	0.45	3.7***
Between person age			β_{01}	0.11 [−0.01, 0.23]	0.07	1.7

*** $p < 0.001$; ** $p < 0.01$; * $p < 0.05$.

process. This result is consistent with the studies performed on the general population prior to the COVID-19 pandemic [27, 28]. Studies have proved that people with high range of anxiety, fear, and loneliness are likely to develop more stress during COVID-19 pandemic [29–31]. The present finding provides empirical support for these claims. This means that controlling the level of anxiety, fear, and loneliness during the time of pandemic will provide a myriad of opportunities in the psychological retrieval even when the COVID-19 stressor is still there. This may include opportunities to physically and mentally distract from the aggravation, build virtual connections, engage in physical activities, and do other household chores during the period of lockdown. However, as unlock period started it brought certain risks but these risks could be managed with various strategies and taking proper precautions [32].

Moreover, the present study reports that the helplessness during the unlock process of COVID-19 was related with poorer psychological retrieval. This finding is similar to the result with the investigation performed prior the COVID-19 pandemic reporting that helplessness is a chief contributor to poor psychological retrieval [33, 34]. Beyond the main effect two findings are worth highlighting regarding helplessness. First similar to emphasizing the role of younger age in reducing helplessness and cross-sectional finding where college employees having older age reported helplessness. Second the relation between helplessness and psychological retrieval among college employees with higher neuroticism, highlights the importance of lower neuroticism in improving the psychological retrieval process during the unlock phases of COVID-19.

Finding of the present study also suggests reduced authenticity is related with poorer psychological retrieval, although effect was comparatively weaker than those among lower neuroticism and helplessness among college employees respectively during COVID-19 unlock phases in India. However, the effect of reduced authenticity on psychological retrieval was not indirectly influenced by helplessness or lower neuroticism. It might be estimated that the reduced authenticity is an alternative for revealing to possibly stressful reportage of COVID-19 which could heighten distress [35]. In the correlational analyzes, higher reduced authenticity was significantly related with greater feeling of helplessness which means that reduced authenticity at the time of pandemic may disturb the employees with ongoing fake news and reports [16, 36]. The researcher also assessed improvement in the psychological retrieval process, helplessness, and reduced authenticity over time. Analyzes of the data in the present study revealed that since unlock-I to unlock-VI in India there was a small improvement in the psychological retrieval process of college employees even when the COVID-19 stressor is still there. The study also reported with the gradual passing of the days during the time of pandemic, there was reported a slight decrease in the feeling of helplessness among college employees [38].

7.5 CONCLUSION

The issue of psychological retrieval is a cause of concern for many during the time of COVID-19 pandemic. The problem of helplessness and reduced authenticity leading to weak autonomy is related to poor psychological retrieval of people during the time of pandemic and unlock process even when the stressor is still there. Those with lower neuroticism characteristics are related to higher psychological retrieval and also alleviated dangerous effect of helplessness. The finding of the present study, particularly the lower neuroticism characteristics may have important implications for psychologists and others in the field. Since the waves and new strain of COVID-19 pandemic may require new norms to be followed as lifestyle such as social distancing, health policies and effectively managing the emotions like anger, anxiety, fear, and depression, especially people of old age. The effectiveness of such condition will promote space for the psychological retrieval by promoting stronger autonomy.

KEYWORDS

- **authenticity**
- **autonomy**
- **COVID-19**
- **global emergency**
- **helplessness**
- **intra-class correlation coefficient**
- **nationwide complete lockdown**
- **neuroticism**
- **psychological retrieval process**

REFERENCES

1. Kandel, N., Chungong, S., Omaar, A., & Xing, J., (2020). Health security capacities in the context of COVID-19 outbreak: An analysis of international health regulations annual report data from 182 countries. *The Lancet, 395*(10229), 1047–1053.
2. Government of India, (2020). *Departmental Order Letter, DO Number- 40-3/2020-DM-1 (A)*. [Press release] https://www.mha.gov.in/sites/default/files/MHAOrder_27122021.pdf (accessed on 5 July 2022).
3. Government of India, (2020). *Press Information Bureau: Lockdown Measures for Containment of COVID-19 Pandemic in the Country to Continue to Remain in Force up to May 3, 2020*. [Press release] https://www.mha.gov.in/sites/default/files/PR_Extensionoflockdown_14042020_0_0.pdf (accessed on 5 July 2022).
4. Sharma, A. J., & Subramanyam, M., (2020). Psychological impact of Covid-19 lockdown in India: Different strokes for different folks. *PLOS One, 308*, 191–199.
5. Roy, D., Tripathy, S., Kar, S. K., Sharma, N., Verma, S. K., & Kaushal, V., (2020). Study of knowledge, attitude, anxiety & perceived mental healthcare need in Indian population during COVID-19 pandemic. *Asian Journal of Psychiatry, 51*, 42–47.
6. Krishna, P. R., Undela, K., Palaksha, S., & Gupta, B. S., (2020). *Knowledge and Beliefs of General Public of India on COVID-19: A Web-based Cross-Sectional Survey, 249*, 797–801. MedRxiv.
7. Pieh, C., Budimir, S., & Probst, T., (2020). Mental health during COVID-19 lockdown: A comparison of Austria and the UK. *SSRN Electronic Journal, 118*, 256–269.
8. Gao, J., Zheng, P., Jia, Y., Chen, H., Mao, Y., Chen, S., & Dai, J., (2020). Mental health problems and social media exposure during COVID-19 outbreak. *PLoS One, 15*(4).
9. Xiao, H., Zhang, Y., Kong, D., Li, S., & Yang, N., (2020). The effects of social support on sleep quality of medical staff treating patients with coronavirus disease 2019 (COVID-19) in January and February 2020 in China. *Medical Science Monitor, 26–38*.
10. Qiu, J., Shen, B., Zhao, M., Wang, Z., Xie, B., & Xu, Y., (2020). A nationwide survey of psychological distress among Chinese people in the COVID-19 epidemic: Implications and policy recommendations. *General Psychiatry*. BMJ Publishing Group.
11. Shankar, A., McMunn, A., Banks, J., & Steptoe, A., (2011). Loneliness, social isolation, and behavioral and biological health indicators in older adults. *Health Psychology, 30*(4), 377–385.
12. Rodriguez, G., Moore, S. J., Neff, R. C., Glass, E. D., Stevenson, T. K., Stinnett, G. S., & Cazares, V. A., (2020). Deficits across multiple behavioral domains align with susceptibility to stress in 129S1/SvImJ mice. *Neurobiology of Stress, 13*.
13. Bliese, P. D., Edwards, J. R., & Sonnentag, S., (2017). Stress and well-being at work: A century of empirical trends reflecting theoretical and societal influences. *Journal of Applied Psychology, 102*(3), 389–402.
14. Ryan, R. M., & Deci, E. L., (2000). The darker and brighter sides of human existence: Basic psychological needs as a unifying concept. *Psychological Inquiry, 11*(4), 319–338.
15. Gilbert, D. T., Wilson, T. D., Pinel, E. C., Blumberg, S. J., & Wheatley, T. P., (1998). Immune neglect: A source of durability bias in affective forecasting. *Journal of Personality and Social Psychology, 75*(3), 617–638.

16. Fried, E. I., Papanikolaou, F., & Epskamp, S., (2021). Mental health and social contact during the COVID-19 pandemic: An ecological momentary assessment study. *Clinical Psychological Science, 13*, 7–18.
17. Horgan, O., & Maclachlan, M., (2009). Disability and rehabilitation psychosocial adjustment to lower-limb amputation: A review psychosocial adjustment to lower-limb amputation: A review. *Taylor & Francis, 26*(14, 15), 837–850.
18. Updegraff, J. A., & Taylor, S. E., (2000). *From Vulnerability to Growth: Positive and Negative Effects of Stressful Life Events, 147*, 155–166.
19. Ryan, R. M., & Deci, E. L., (2006). Self-regulation and the problem of human autonomy: Does psychology need choice, self-determination, and will?. *Journal of Personality, 74*(6), 1557–1586.
20. Gau, S. F., (2000). Neuroticism and sleep-related problems in adolescence. *Sleep, 23*(4), 495–502.
21. Aarts, H., Dijksterhuis, A., & De Vries, P., (2001). On the psychology of drinking: Being thirsty and perceptually ready. *British Journal of Psychology, 92*(4), 631–642.
22. Seibt, B., Häfner, M., & Deutsch, R., (2007). Prepared to eat: How immediate affective and motivational responses to food cues are influenced by food deprivation. *European Journal of Social Psychology, 37*(2), 359–379.
23. Strack, F., & Deutsch, R., (2004). Reflective and impulsive determinants of social behavior. *Personality and Social Psychology Review, 8*(3), 220–247.
24. Radel, R., Pelletier, L. G., Sarrazin, P., & Milyavskaya, M., (2011). Restoration process of the need for autonomy: The early alarm stage. *Journal of Personality and Social Psychology, 101*(5), 919–934.
25. Curran, P. J., & Bauer, D. J., (2011). The disaggregation of within-person and between-person effects in longitudinal models of change. *Annual Review of Psychology, 62*, 583–619.
26. Ormel, J., Jeronimus, B. F., Kotov, R., Riese, H., Bos, E. H., Hankin, B., & Oldehinkel, A. J., (2013). Neuroticism and common mental disorders: Meaning and utility of a complex relationship. *Clinical Psychology Review, 33*(5), 686–697.
27. Schneider, T. R., Rench, T. A., Lyons, J. B., & Riffle, R. R., (2012). The influence of neuroticism, extraversion and openness on stress responses. *Stress and Health, 28*(2), 102–110.
28. Sher, L., (2020). The impact of the COVID-19 pandemic on suicide rates. *QJM: An International Journal of Medicine, 113*(10), 707–712.
29. Newby, J. M., O'Moore, K., Tang, S., Christensen, H., & Faasse, K., (2020). Acute mental health responses during the COVID-19 pandemic in Australia. *PloS One, 15*(7), e0236562.
30. Lebel, C., MacKinnon, A., Bagshawe, M., Tomfohr-Madsen, L., & Giesbrecht, G., (2020). Elevated depression and anxiety symptoms among pregnant individuals during the COVID-19 pandemic. *Journal of Affective Disorders, 277*, 5–13.
31. Freeman, S., & Eykelbosh, A., (2020). COVID-19 and outdoor safety: Considerations for use of outdoor recreational spaces. *National Collaborating Centre for Environmental Health, 14*, 1–15.
32. Son, C., Hegde, S., Smith, A., Wang, X., & Sasangohar, F., (2020). Effects of COVID-19 on college students' mental health in the United States: Interview survey study. *Journal of Medical Internet Research*. JMIR Publications.

33. Zvolensky, M. J., Garey, L., Rogers, A. H., Schmidt, N. B., Vujanovic, A. A., Storch, E. A., & O'Cleirigh, C., (2020). Psychological, addictive, and health behavior implications of the COVID-19 pandemic. *Behavior Research and Therapy, 134*, 103715.
34. Coburn, S. S., Gonzales, N. A., Luecken, L. J., & Crnic, K. A., (2016). Multiple domains of stress predict postpartum depressive symptoms in low-income Mexican American women: The moderating effect of social support. *Archives of Women's Mental Health, 19*(6), 1009–1018.
35. Dubey, S., Biswas, P., Ghosh, R., Chatterjee, S., Dubey, M. J., Chatterjee, S., & Lavie, C. J., (2020). Psychosocial impact of COVID-19. Diabetes & metabolic syndrome. *Diabetes & Metabolic Syndrome: Clinical Research & Reviews, 14*(5), 779–788.
36. Gradoń, K., (2020). Crime in the time of the plague: Fake news pandemic and the challenges to law enforcement and intelligence community. *Society Register, 4*(2), 133–148.

CHAPTER 8

Depression, Anxiety, and Stress of Coronavirus-Infected People of Kashmir in Relation to Psychological Hardiness

SHABIR AHMAD MALIK

Research Scholar, Department TT and NFE, IASE, Faculty of Education, Jamia Millia Islamia University, New Delhi, India, E-mail: malikshabir622@gmail.com

ABSTRACT

A coronavirus outbreak and its associated measures have caused mental health problems such as stress, worry, sadness, insomnia, denial, wrath, and panic throughout the world due to their rapid spread and their resulting actions. People living in Kashmir who have been infected with a coronavirus were assessed for sadness, anxiety, and stress, and for psychological resilience. A total of 120 participants were selected through purposive sampling from two districts (Ganderbal and Srinagar) of Kashmir Valley. Depression anxiety stress scale (DASS-21) and psychological hardiness scale were used for data collection. Results revealed that 66%, 17%, 10%, and 5% were normal to extremely severe depression. 45%, 21%, 15%, 10%, and 8% of the participants were normal to extremely severe anxiety. 35%, 25%, 19%, 11%, and 10% of the participants were normal to extremely severe stress. Results also indicated that 15% of people have a low level of psychological hardiness, 71% have an average level, and 15% have higher psychological hardiness. Psychological hardiness and DASS-21 showed a highly significant association (depression, anxiety,

---

Emotional Intelligence for Leadership Effectiveness: Management Opportunities and Challenges During Times of Crisis. Mubashir Majid Baba, Chitra Krishnan, & Fatma Nasser Al-Harthy (Eds.)
© 2023 Apple Academic Press, Inc. Co-published with CRC Press (Taylor & Francis)

and stress). As a result, the study found that patients who are infected with the COVID-19 coronavirus require proper counseling to minimize its psychological impact. An increase in psychological hardiness would also reduce depression, anxiety, and stress in coronavirus-infected people.

8.1 INTRODUCTION

People's mental health suffers when they are confined to a certain type of environment. The psychological effects of lockdown, isolation, and quarantine to contain pandemics have been studied in a number of studies. Among quarantined parents and children, 25% had posttraumatic stress disorder, and 30% had it [1]. While in isolation in Korea, 7.6% of 1,656 patients reported anxiety symptoms and 16.6% displayed rage [2]. When SARS (severe acute respiratory syndrome) broke out in Canada in 2003, similar consequences were observed [3]. As a result of the lockdown, mental illness is on the rise, according to several sources. Fast the entire planet has been paralyzed by this disease's terrible outbreak. According to the World Health Organization, over 200 nations have reported corona-positive cases [4]. The first coronavirus pandemic case was reported in India on January 30, 2020. The virus was believed to have originated from China. As well as the state of Jammu and Kashmir, the union territory was also impacted. In Jammu and Kashmir, this is the first time that a fatal virus has been discovered [5]. Jammu and Kashmir has had 80,474 cases of the disease, including 13,712 active cases, 65,496 successful recoveries, and 1,268 deaths, as of October 2, 2020 [6]. People began to worry about the coronavirus pandemic because of its extremely high infection rate and comparable high mortality rate. Stress, concern, depressive symptoms, insomnia, denial, fury, and fear are only a few of the mental health challenges produced by the rapid spread of this disease and associated measures worldwide [7–11]. During the lockdown, anxiety, depression, tension, and other mental health concerns were frequent, according to the psychiatric society of Goa [12]. There has been a 20% increase in the number of patients who suffer from mental illness [13].

Due to the coronavirus pandemic, life is full of thrilling challenges, terrible situations, anxiety, fears, and emotional problems. COVID-19 has forced people to adopt specific personality traits in order to function effectively. People who have high levels of psychiatric hardiness,

Depression, Anxiety, and Stress of Coronavirus-Infected 123

according to study, are better able to handle such situations. Psychological hardiness refers to a person's ability to effectively deal with stress. It is the process of adapting well to adversity, such as a trauma or a failure, when faced with it. A person is more likely to be able to deal with changes and problems in their lives with a sense of humor and resilience, which has a positive impact on their health-preventing behavior. Pressure, worry, and hopelessness are the norm for most of us in today's world of business. Suffering from melancholy and accepting the death of a loved one are necessary for survival. As a result, it is used to assess a person's propensity to form a bond with themselves and their environment. The purpose of the creative process is not rigidity or stress tolerance, but rather the ability to properly navigate difficult situations and stressful situations. In other words, it does not indicate that you are careless. It only means that you are capable of detecting the conditions around you and making your own choices. Early research identified a personality structure consisting of three interconnected dispositions: commitment, control, and challenge, as the basis for a person's ability to cope with stressful situations. One's propensity to engage in life's activities fully and with dedication is known as the commitment disposition. By putting in their hard effort, an individual tends to assume that they can influence an especially stressful scenario. Lastly, the challenge disposition is defined as "the conviction that life transitions are opportunities for personal progress" [14]. In other words, it is the ability to see the shift as an opportunity for growth and progress. Stressors can be tolerated without resulting in mental illness or persistently negative mood [15, 16]. Development and environmental stresses can be reduced by fostering mental and physical health growth in emerging countries with these attitudes [17, 18]. As a result of a pattern of attitudes and talents, you can turn potentially disastrous events into growth possibilities [19].

8.2 OBJECTIVES

- To determine the level of depression, anxiety, stress, and psychological hardiness among coronavirus-infected people of Kashmir;
- To find out the relationship of psychological hardiness with depression, anxiety, and stress among coronavirus-infected people of Kashmir.

124   *Emotional Intelligence for Leadership Effectiveness*

8.3 METHOD

8.3.1 PARTICIPANTS

Using a purposive sampling strategy, 120 coronavirus-infected persons were recruited in Ganderbal and Srinagar, two districts of Kashmir Valley. Participants were made aware of the study's nature and purpose, as well as its voluntary nature.

8.3.2 MEASURES

8.3.2.1 DEPRESSION ANXIETY STRESS SCALE (DASS-21)

This study employed the depression anxiety stress scale (DASS-21). It gauges three negative states and has 21 components. As a result of the study, participants completed three subscales each having seven items that measured depression, anxiety, and stress levels. As a result, responses ranged from 0 to 3, with zero meaning that it did not apply to me; one meaning that it did apply to me in some way; two meaning that it did apply to me in a significant way or a good part of the time; and three meaning that it did apply to me very much or most of the time. 63 (21 multiplied by three) is the maximum possible score (21×0). It indicates a higher level of depressive and anxious symptoms.

8.3.2.2 PSYCHOLOGICAL HARDINESS SCALE

In this study, the researcher himself established the Psychological Hardiness Scale. Measures three dimensions (Commitment, Control, and Challenge) and consists of 45 items. There is a five-point scale for grading the responses. Positive items are scored as follows: strongly agree receives a score of 5, agree receives a score of 4, neither agree nor disagree receives a score of 3, disagree receives a score of 2, and strongly disagree receives a score of 1. As for the negative items, highly agree receives a score of 1, agree gets a score of 2, neither agree nor disagree receives a score of 3, disagree gets a score of 4, and strongly disagree receives the highest possible score of 5. A higher score indicates better mental toughness. 225 (45×5) is the maximum psychological toughness score, while 45 is the least score in this category (45×1 marks).

Depression, Anxiety, and Stress of Coronavirus-Infected 125

8.3.3 STATISTICAL TREATMENT

For achieving the objectives of the present study, the collected data was analyzed by using the appropriate statistical techniques with the help of SPSS-20.

8.4 RESULTS AND INTERPRETATION

The results and their interpretation have been presented in Tables 8.1–8.5. Table 8.1 indicates that 66% of coronavirus-infected people have a normal level of depression, 17% have a mild level of depression, 10% have a moderate level of depression, 5% have a severe level of depression, and 0% have extremely severe depression.

TABLE 8.1 Level of Depression Among Coronavirus Infected People of Kashmir (n = 120)

Depression					Total
Normal (0–9)	Mild Depression (10–13)	Moderate Depression (14–20)	Severe Depression (21–27)	Extremely Severe Depression (28+)	
80 (66%)	20 (17%)	13 (10%)	07 (5%)	0%	120

Table 8.2 indicated that 45% of coronavirus-infected people have a normal level of anxiety, 21% have a moderate level of anxiety, 15% have a moderate level of anxiety, 10% have a severe level of anxiety, and 8% have extremely severe anxiety.

TABLE 8.2 Level of Anxiety Among Coronavirus Infected People of Kashmir (n = 120)

Anxiety					Total
Normal (0–7)	Mild Anxiety (8–9)	Moderate Anxiety (10–14)	Severe Anxiety (15–19)	Extremely Severe Anxiety (20+)	
54 (45%)	26 (21%)	18 (15%)	12 (10%)	10 (8%)	120

Table 8.3 indicates that 35% of coronavirus-infected people have a normal level of stress, 25% of people have a mild level of stress, 19% have a moderate stress level, 11% people have severe stress, and 10% people have extremely severe stress.

126　　　　　　　　*Emotional Intelligence for Leadership Effectiveness*

TABLE 8.3　Level of Stress Among Coronavirus Infected People of Kashmir (n = 120)

Stress					Total
Normal (0–14)	**Mild Stress (15–18)**	**Moderate Stress (19–25)**	**Severe Stress (26–33)**	**Extremely Severe Stress (34+)**	
42 (35%)	29 (25%)	23 (19%)	14 (11%)	12 (10%)	120

Table 8.4 indicates that 15% of coronavirus-infected people have a low level of commitment, 70% have an average level, and 14% were having a higher level of commitment; 13% of people have a low level of control, 74% have an average level, and 12% were having a higher level of control; 17% coronavirus infected people have a low level of challenge, 67% have an average level, and 15% were having a higher level of challenge; 15% people have a low level of psychological hardiness, 71% have an average level, and 15% people were having a higher level of psychological hardiness.

TABLE 8.4　Level of Psychological Hardiness Among Coronavirus Infected People of Kashmir (n = 120)

	Levels					
	Low	**%**	**Average**	**%**	**High**	**%**
Commitment	18	15	85	70	17	14
Control	16	13	89	74	15	12
Challenge	21	17	80	67	19	15
Total psychological hardiness	17	15	87	71	16	15

Table 8.5 results revealed a negative correlation between psychological hardiness and DASS-21 (depression, anxiety, and stress). The intensity of the correlation is ($r = -0.59$) and found statistically significant at 0.01 level. This suggests that higher would be the psychological hardiness, lower would be the depression, anxiety, and stress among coronavirus-infected people.

8.5　DISCUSSION

Coronavirus-infected people in Kashmir were assessed for sadness, anxiety, tension, and psychological toughness. There were 66% of people with a normal depression level, 17% who had mild depression, 10% who had moderate depression, 5% who had a severe level of depression, and

Depression, Anxiety, and Stress of Coronavirus-Infected 127

0% who had extremely severe depression, according to the results in Table 8.1.

TABLE 8.5 Correlation Coefficient of Psychological Hardiness with Depression, Anxiety, and Stress Among Coronavirus Infected People of Kashmir

Variable	Correlation
Psychological hardiness	–0.59
Depression, anxiety, and stress	

Significant at the 0.01 level.

On the basis of the data in Tables 8.2 and 8.3, 45% of patients infected with coronavirus have a normal level of anxiety, 21% have a mild level of anxiety, 15% have a moderate level of anxiety, and 10% have a severe level of anxiety. Around 35% of coronavirus-infected patients have a normal level of stress, 29% have mild stress, 23% have moderate stress, 11% have severe stress, and 10% have extremely severe stress.

Table 8.4 shows that 15% of coronavirus-infected people have a low degree of dedication, 70% have a medium level of commitment, and 14% have a high level of commitment; 13% of persons have a low degree of control, 74% have a medium level, and 12% have a high level of control; The challenge level for 17% of coronavirus-infected patients is low, 67% is average, and 15% is high. The psychological hardiness of 15% of persons is low, 71% is average and 14% is high. The results also revealed a negative correlation between psychological hardiness and DASS-21 (depression, anxiety, and stress). The intensity of the correlation is ($r = -0.59$) and found statistically significant at 0.01 level. This suggests that higher would-be psychological hardiness lower would be the depression, anxiety, and stress among coronavirus-infected people. In previous studies, a negative association between hardiness and stress has been found [20, 21]. Workplace mental stress and psychological toughness are negatively correlated [22]. Stress reduces the physical erosion of psychological toughness [21]. Psychological hardiness is a form of defense against stress and its repercussions [23]. In challenging situations, psychological hardiness might help people to improve their health and performance [24].

The Jammu Kashmir government and mental health specialists must pay immediate attention to the mental health of inhabitants. People's mental health has been negatively impacted by the coronavirus outbreak and the ongoing lockdown. As part of their research, scientists are aiming to discover the coronavirus' genetic makeup as well as its epidemiological characteristics and clinical symptoms around the world. Unfortunately, the psychological impact of the coronavirus pandemic is overlooked. Stakeholders should be alerted so that they can act quickly [25].

8.5.1 LIMITATIONS OF THE STUDY

Due to the small sample size results of the present study cannot be genuinely generalized. Another limitation is that several variables were not measured that may influence psychological hardiness with depression, anxiety, and stress (DASS-21). Future studies could be conducted by considering other variables. Besides, it is suggested that future researchers may undertake such studies with a large sample so that the results obtained may be more reliable.

ACKNOWLEDGMENTS

Many thanks to Dr Eram Nasir, Assistant Professor in the department of teacher training and non-formal education (NFE), Jamia Millia Islamia University, New Delhi, India, to carry out this research work. After that, my special thanks to the coronavirus-infected people of Kashmir for their participation in this study.

KEYWORDS

- anxiety
- coronavirus infected people
- COVID-19
- depression
- depression anxiety stress scale
- psychological hardiness
- stress

Depression, Anxiety, and Stress of Coronavirus-Infected

REFERENCES

1. Sprang, G., & Silman, M., (2013). Posttraumatic stress disorder in parents and youth after health-related disasters. *Disaster Medicine and Public Health Preparedness, 7*(1), 105–110.
2. Jeong, H., Yim, H. W., Song, Y. J., Ki, M., Min, J. A., Cho, J., & Chae, J. H., (2016). Mental health status of people isolated due to middle east respiratory syndrome. *Epidemiology and Health, 38,* Article e2016048.
3. Reynolds, D. L., Garay, J. R., Deamond, S. L., Moran, M. K., Gold, W., & Styra, R., (2008). Understanding compliance and psychological impact of the SARS quarantine experience. *Epidemiology & Infection, 136*(7), 997–1007.
4. Worldometer, (2020). *Countries Where COVID-19 has Spread.* https://www.worldometers.info/coronavirus/countries-where-coronavirus-has-spread/ (accessed on 5 July 2022).
5. J&K, Health & Medical Education Department (2020). *One Case of a Patient Admitted in Isolation at GMC Tested Positive for Coronavirus, the Patient Had Travel History to Iran, And Second Patients Sample Being Sent for Retest.* jkhealth.org. Retrieved from: https://twitter.com/HealthMedicalE1/status/1236894502670839808 (accessed on 5 July 2022).
6. MoHFW-GOI (2020). *Ministry of Health and Family Welfare.*
7. Brooks, S. K., Webster, R. K., Smith, L. E., Woodland, L., Wessely, S., Greenberg, N., & Rubin, G. J., (2020). The psychological impact of quarantine and how to reduce it: A rapid review of the evidence. *Lancet, 395,* 912–920.
8. Galea, S., Merchant, R. M., & Lurie, N., (2020). The mental health consequences of COVID-19 and physical distancing: The need for prevention and early intervention. *JAMA Internal Medicine,* E1, E2. 10.1001/jamainternmed.2020.1562.
9. Qiu, J., Shen, B., Zhao, M., Wang, Z., Xie, B., & Xu, Y., (2020). A nationwide survey of psychological distress 331 among Chinese people in the COVID-19 epidemic: Implications and policy recommendations. *General of Psychiatry, 33,* 19 21.
10. Torales, J., O'Higgins, M., Castaldelli-Maia, J. M., & Ventriglio, A., (2020). The outbreak of COVID-19 coronavirus and its impact on global mental health. *International Journal of Social Psychiatry,* 1–4. 10.1177/0020764020915212.
11. Xiang, Y. T., Yang, Y., Li, W., Zhang, L., Zhang, Q., & Cheung, T., (2020). Timely mental health care for the 2019 novel coronavirus outbreak is urgently needed. *The Lancet Psychiatry, 7,* 228, 229.
12. PTI. (2020). *Goa: Coronavirus Lockdown Triggers a Rise in Mental Health Issues.* https://www.deccanherald.com/national/west/goa-coronavirus-lockdown-triggers-riseinmental- health-issues-823707.html (accessed on 5 July 2022).
13. Lolwal, M., (2020). *20% Increase in Patients with Mental Illness Since the Coronavirus Outbreak: Survey.*
14. Kobasa, S. C., (1979). Stressful life events, personality, and health–Inquiry into hardiness. *Journal of Personality and Social Psychology, 37*(1), 1–11.
15. Neill, J., (2009). Psychological hardiness, positive emotions, and successful adaptation to stress in later life. *Journal of Personality and Social Psychology, 51*(3), 478–509.
16. Reivich, K., & Shatte, A., (2008). *The Resilience Factor: 7 Keys to Finding Your Inner Strength and Overcoming Life's Hurdles.* N.J.: Broadway Random House.

17. Antonovsky, A., (1979). *Health, Stress, and Coping*. San Francisco: Jossey-Bass.
18. Maddi, S. R., (2002). The story of hardiness: Twenty years of theorizing, research, and practice. *Consulting Psychology Journal, 54*, 173–185.
19. Maddi, S. R., & Khoshaba, D. M., (2007). The relevance of hardiness assessment and training to the military context. *Journal of Military Psychology, 19*(1), 61–70.
20. Sheppard, J. A., & Kashani, J. H., (1999). The relationship of hardiness, gender, and stress to health outcomes in adolescents. *Journal of Personality, 59*(4), 747–767.
21. Majidian, F., (2004). *The Relationship of Self-Efficacy Beliefs and Hardiness with Direct Job Stress*. The MA thesis. The psychological department of Allame Tabatabai University.
22. Izkian, S. S., (2001). *A Comparison of Violence and Hardiness*. M.A. thesis General Psychology, Tarbiat Moalem University.
23. Sadaghiani, N. S. K., (2011). The role of hardiness in decreasing the stressors and biological, cognitive, and mental reactions. *Procedia-Social and Behavioral Sciences, 30*, 2427–2430.
24. Maddi, S. R., (1999). Comments on trends in hardiness research and theorizing. *Consulting Psychology Journal: Practice & Research, 51*, 67–71.
25. World Health Organization, (2020). *Mental Health and Psychological Resilience During the COVID-19 Pandemic.*

CHAPTER 9

Self-Management During the COVID-19 Crises

REHANA AMIN and MOHAMMAD MAQBOOL DAR

Faculty of Department of Psychiatry, Government Medical College Srinagar (IMHANS–Kashmir), Jammu and Kashmir, India, E-mail: drrehanaamin2@gmail.com (R. Amin)

ABSTRACT

Since the COVID-19 pandemic hit the world, a number of health monitoring and diagnostic systems have been proposed for e-health care ranging from simple wearable type of device or sensor to complex implantable sensors. Despite of all this there are several challenges both in the technological/clinical and managerial. Since the COVID-19 pandemic forces the human race to remain isolated and Quarantine at the appropriate places. In such scenarios the technological and scientific advances may not work. Therefore, in this work, some self-strategies have been proposed for mitigation of COVID-19 crisis. The proposed strategies are in tune with the clinical and scientifically principles. The appropriate materials and methods have been used to validate the results.

9.1 INTRODUCTION

A situation during which an individual physical, mental, and social well-being is at risk is known as a crisis situation [1]. A crises situation like the pandemic of novel coronavirus (nCoV) disease has hit every corner of the

---

Emotional Intelligence for Leadership Effectiveness: Management Opportunities and Challenges During Times of Crisis. Mubashir Majid Baba, Chitra Krishnan, & Fatma Nasser Al-Harthy (Eds.)
© 2023 Apple Academic Press, Inc. Co-published with CRC Press (Taylor & Francis)

world by November 2019 and has affected almost everyone in one way or another. Some got infected, while most were under psychological burden during this pandemic [2]. COVID-19, which catches mankind off guard and therefore not having an effective treatment brings manifold troubles to people because it not only affects their health but also sweeps other important domains of life. The pandemics like COVID-19, which are beyond our control, can induce fear and uncertainty in every individual [3]. The response to any event depends on the nature of the event, attitude, and perception of people towards an event and more importantly a person's psychological predisposition [4]. To negotiate and counter a global crisis like a pandemic, self-management is extremely important for an all-inclusive individual.

While growing older we face different challenges and risks in life. Modern life temptations like a sedentary lifestyle, unhealthy diets, substance abuse, crowded working places, and an unfavorable environment predispose an individual more to health-related problems [5]. The habit of practicing self-care strategies substantially reduces the risk of health-related problems thereby helping in dealing with a crisis-like situation. Self-care is defined as an approach towards the self that entails inculcating and exercising things that inevitably enhance the physical and mental health of a person. Self-care a part of routine life is disturbed at times of crisis [6].

More research is done on self-care related to non-communicable diseases, and little evidence is related to self-care during pandemics like COVID-19. However, the same self-care management plan works and can prove beneficial during these crises. At present self-care is a priority but is lacking in actuality due to knowledge gaps. So, the aim is to address these knowledge gaps by generating a self-care intervention plan which contributes to effective well-being, lower morbidity and improves the health status of a nation [7]. It is not possible to form a self-care plan during times of crisis but to have a routine self-care strategy in hand can prove beneficial in any challenging situation to cope with and maintain well-being [8].

9.2 MATERIAL AND METHODS

The data was collected by conducting a systematic review of related articles published till February 2021. The search engines used were Google

Scholar, PubMed, Medline, EMBASE, and Wiki.com. The articles were searched by typing self-care, self-management, pandemic, crises, COVID-19, Self-care methods, self-care strategy and health education program as a search query in search engines. All the related articles were thoroughly reviewed and relevant material was collected.

9.3 RESULTS AND DISCUSSION

9.3.1 SELF-CARE CHALLENGES

Self-care challenges during pandemics are the same as seen in the care of non-communicable diseases. Worldwide various factors have been found which can motivate an individual to take an interest in self-care. Like people lack the knowledge of basic health information, which is the major challenge contributing to poor self-care. A poor health care education creates a delay in early detection of health-related problems which in turn increases morbidity in future [9]. Poor access to health care services or a shortage of resources is also a big barrier affecting the self-care of an individual [10]. People from poor socioeconomic classes also lack resources as well as knowledge so are unable to meet their health care needs therefore engage less in self-care activities [11, 12]. The patients suffering from mental illnesses do have impaired cognitive functions, which lead to socio-occupational dysfunction and failure to follow healthy lifestyle behaviors, thus compromising the self-care process in them too [13]. The indwelling premorbid personality, psycho-social factors, interpersonal factors like age, lack of prior experience, multiple medical comorbidities, and influence of others are also challenging that we face in the self-management process [14–16].

Even when exposed to the same challenges, different people face different problems and have different self-care needs. The self-care needs vary from basic physical, psychosocial, monetary, emotional, spiritual, supportive, and professional needs [17]. Self-care policy programs have been recommended. The policies which are universal in their application and focus on supporting positive behaviors in healthy people are more appropriate, better accepted, and useful than those which penalize sick people or target individuals [18].

9.3.2 SELF-CARE STRATEGY

The cycle of activity behavior of maintenance, monitoring, and managing of self is known as self-care [8]. The care of already existing well-being, promotion of health, prevention of diseases is known as self-care maintenance. It is the measure to improve physical and emotional health by adapting to a healthy lifestyle [19]. Self-care monitoring is to know the normalcy of self and recognize changes at the earliest to prevent disability [20]. Self-care management represents a promising strategy for solving problems by identifying challenges. Various self-management strategies can be employed to deal with grave precedented and unprecedented crises. The self-management strategies provide a useful framework for an individual to respond to and effectively combat any crisis.

9.3.2.1 PSYCHOEDUCATION

The first step to deal with this particular crisis is psychoeducation which means an individual must know the general information about COVID-19 infection, its symptoms, modes of transmission, and its preventive measures so that a timely strategy is followed to avoid the spread of the disease. WHO has provided us with some simple precautionary measures to stay safe. We should stick to the basic preventive measures like regular hand washing, use of triple-layer masks, avoiding unnecessary outings or social gatherings, maintaining a physical distance of at least 1 meter, and adhering to good respiratory hygiene to prevent the spread of infection. Psychoeducation not only helps us to curtail the spread of infection but also decreases the fear of getting the infection [21].

9.3.2.2 HYGIENE

We should make provisions for adequate ventilation to ensure the fresh and crisp air in living rooms and also at a working place to boost the self-management program. Make it a habit to clean and disinfect the frequently touched surfaces. The timely self-quarantine measures must be observed with utmost vehemence if one comes in contact with a COVID-19 case to make sure the non-transmission of the virus to others. To maintain oneself it is recommended to seek medical attention immediately once symptoms of COVID-19 are suspected [22].

9.3.2.3 PHYSICAL ACTIVITY

The matter of concern during the COVID-19 pandemic has been the activity levels which greatly reduced due to the lockdown imposed to curtail the spread of infection. The famous adage runs that a healthy mind resides in a healthy body and we can have a healthy body only by practicing a healthy lifestyle. Daily exercise or YOGA (your objectives, guidelines, and assessment) at home not only improves our overall physical health but also significantly relieves us of mental stress. A study revealed that people who had a healthy lifestyle before the pandemic adhere more to coronavirus protective behaviors which not only have direct benefits but also prepare an individual for the next global health crisis [23].

9.3.2.4 HEALTHY HABITS

Increased time at home and COVID-19 fear push people towards over-eating. The stress may provoke a craving for certain foods and also disturbs sleep. We should observe proper sleep hygiene and eat healthy foods to replenish and repair the body. Good food and proper rest facilitate the regeneration and detoxification of the body. People should completely refrain from substance abuse because such indulgences can aggravate the existing psychological crisis brought about by the pandemic. One should observe a strict regimen for the ailments which linger throughout life to prevent exacerbation [24].

9.3.2.5 AVOID STRESS

During the pandemic, a common stress-inducing factor has been the persistent updates about COVID-19 given on different social media platforms. TV programs like comedy shows, religious programs, health programs, or programs of choice act as real stress busters. The quality time spent in indoor games, household chores, family discussions, indoor hobbies and working from home significantly reduce the feeling of boredom arising out of the non-movement protocol [25]. Religion which has been the refuge of mankind since the dawn of human civilization comes in handy even in times of such crisis. One should follow religious teachings and listen to religion, which no other thing can provide [26]. People should adopt a positive approach to life. One should bear in mind that everything good or bad is fickle or temporary in this world. With the above-mentioned mindset, one should focus on immediate needs and put the remote and less important things on the back burner to relieve oneself of undue stress [27].

9.3.2.6 EFFECTIVE INTERVENTION WITH HEALTH CARE PROVIDERS

For timely management of symptoms and disability, the appropriate and wise use of modern digital technology can be of tremendous help in the crisis-ridden period. At a time when visiting a doctor in person can have undesirable consequences, the internet can significantly come in handy by facilitating virtual consultations [28].

9.3.2.7 MAINTAINING ROLE AND RELATIONSHIPS

One can maintain healthy relationships and friendships even by doing away with the real in-person meetings by availing various social media platforms The aforementioned mobile phone network can be used in blissful conference calls among peers and relatives wherein they can indulge in humorous and spirit boosting talks and also share worries if any amongst themselves. The restrictions on the movement and free flow of people by imposed lockdowns can take a heavy toll on the education and career of students, but this effect can somehow be mitigated by arranging online classes and healthy virtual interactions amongst the students and teacher [29].

9.3.2.8 STIGMA

The COVID-19 carries a significant risk of stigma to frontline workers as well as to those who were infected with the infection. As per WHO "All efforts must be taken to scientifically destigmatize COVID-19 instead of

Self-Management During the COVID-19 Crises 137

statutory sermons by lawmakers." "Proper health education targeting the public appears to be the most effective method to prevent social harassments of both healthcare workers and COVID-19 survivors." Avoiding stigma related to pandemics makes it easy to contain the spread of infection [30].

9.3.2.9 PRACTICAL ASSISTANCE

Exposure to any crisis situation is often accompanied by hopelessness in highly vulnerable people. In such situations, we can improve the emotional and mental health of ourselves as well as others by helping the disadvantaged. By listening to their concerns and meeting their basic needs, one can help the warriors and can improve their own emotional, physical, moral, and social well-being. Therefore, assisting the survivor with problem-solving helps in managing self during crises [31].

9.3.2.10 MEDITATION AND MINDFULNESS

Meditation is a technique that would help to deliberately shift attention from routine to present to achieve the stable mental and physical state of an individual, while mindfulness generates self-concentration so that we can monitor our health and detect changes in the body [32].

9.3.2.11 HEALTHY CONDUCIVE WORKING ENVIRONMENT

Working in a favorable healthy environment also promotes healthy functioning during times of crisis. It proves to be an important factor in reducing stress in grave situations [33].

9.4 CONCLUSION

COVID-19 pandemic proved to be an individual as well as a global crisis. Where in both situations, an individual itself needs to be courageous and a front runner for providing numerous solutions to cope with the crises. During times of crisis, the maintenance of the competency of an individual is important for integrity. The role of daily self-assessment and self-management prepares an individual to fight back and limit the damage caused during any type of crisis. Various self-management strategies already discussed provide a useful framework for an individual to respond to and effectively combat any crisis. These strategies are not confined and applicable to any particular crisis but are universal and can be effectively applied to any crisis's situation like COVID-19. These strategies not only keep the resulting anxiety at bay but will also develop a positive outlook and optimism among the people and therefore enable them to cope with the problem at hand with sheer efficacy.

KEYWORDS

- **COVID-19**
- **pandemic**
- **psychoeducation**
- **self-care management**
- **strategy**

REFERENCES

1. Bundy, J., Pfarrer, Short, C. E., & Coombs, W. T., (2017). Crises and crisis management: Integration, interpretation, and research development. *Journal of Management, 43*(6), 1661–1692. doi: 10.1177/0149206316680030.
2. Varshney, M., Parel, J. T., Raizada, N., & Sarin, S. K., (2020). Initial psychological impact of COVID-19 and its correlates in Indian Community: An online (FEEL-COVID) survey. *PLoS One, 15*(5), e0233874. doi: 10.1371/journal.pone.0233874.
3. Kadidiatou, F. N., Latif, F., Sarfraz, S., et al., (2020). Fear and agony of the pandemic leading to stress and mental illness: An emerging crisis in the novel coronavirus (COVID-19) outbreak. *Psychiatry Research*, 113230. https://doi.org/10.1016/j.psychres.2020.113230.
4. Betty, P., Brian, H. J., Carol, S. N., & James, L., (2008). Regens youth's reactions to disasters and the factors that influence them. *Prev. Res., 15*(3), 3–6. doi: 10.1901/jaba.2008.15-3.
5. Dariush, D. F., (2015). Impact of lifestyle on health. *Iran J. Public Health, 44*(11), 1442–1444.

6. Kollack, I., (2006). The concept of self care. In: Kim, H. S., & Kollak, I., (eds.), *Nursing Theories: Conceptual and Philosophical Foundations* (pp. 45–51). Springer Publishing Company. ISBN 978-0-8261-4006-7.

7. Barbara, R., et al., (2019). Self-care research: Where are we now? Where are we going? *Int. J. Nurs. Stud.*, 103402. doi: 10.1016/j.ijnurstu.2019.103402.

8. Matarese, M., Clari, M., De Marinis, M. G., Barbaranelli, C., Ivziku, D., Piredda, M., & Riegel, B., (2019). The self-care in chronic obstructive pulmonary disease inventory: Development and psychometric evaluation. *Eval. Health. Prof.*, 163278719856660.

9. Wallace, A. S., Carlson, J. R., Malone, R. M., Joyner, J., & DeWalt, D. A., (2010). The influence of literacy on patient-reported experiences of diabetes self-management support. *Nursing Research, 59*(5), 356363. doi: 10.1097/nnr.0b013e3181ef3025.

10. DeVoe, J. E., Baaez, A., Angier, H., Krois, L., Edlund, C., & Carney, P. A., (2007). Insurance + access ≠ health care: A typology of barriers to health care access to low-income families. *Annals of Family Medicine, 5*(6), 511–518. doi: 10.1370/afm.748.

11. Parikh, P. B., et al., (2014). The impact of financial barriers on access to care, quality of care and vascular morbidity among patients with diabetes and coronary heart disease. *Journal of General Internal Medicine, 29*(1), 76–81. doi: 10.1007/s11606-013-2635-6.

12. Whitaker, K. L., Scott, S. E., & Wardle, J., (2015). Applying symptom appraisal models to understand sociodemographic differences in responses to possible cancer symptoms: A research agenda. *Br. J. Cancer, 112*(Suppl 1), S27–34.

13. Zizolfi, D., Poloni, N., Caselli, I., Ielmini, M., Lucca, G., Diurni, M., Cavallini, G., & Callegari, C., (2019). Resilience and recovery style: A retrospective study on associations among personal resources, symptoms, neurocognition, quality of life and psychosocial functioning in psychotic patients. *Psychol. Res. Behav. Manag., 12*, 385–395.

14. Skinner, T. C., Bruce, D. G., Davis, T. M., & Davis, W. A., (2014). Personality traits, self-care behaviors and glycaemic control in type 2 diabetes: The Fremantle diabetes study phase II. *Diabet. Med., 31*(4), 487–492.

15. Arnault, D. S., (2018). Defining and theorizing about culture: The evolution of the cultural determinants of help-seeking, revised. *Nurs. Res., 67*(2), 161–168.

16. Changizi, M., & Kaveh, M. H., (2017). Effectiveness of the mHealth technology in the improvement of healthy behaviors in an elderly population-a systematic review. *Mhealth, 3*, 51.

17. Amanda, M. C., Linda, L. P., Barbara, H., & Victoria, S., (2006). Self-care needs of caregivers dealing with stroke. *Journal of Neuroscience Nursing, 38*, 1.

18. Fiona, M., Charlotte, P., William, E. H., & Nicky, B., (2011). Self-care in people with long term health problems: A community-based survey. *BMC Fam Pract., 12*, 53. doi: 10.1186/1471-2296-12-53.

19. Hui-Wan, C., Chi-Wen, K., Wei-Shiang, L., & Yue-Cune, C., (2019). Factors affecting self-care maintenance and management in patients with heart failure testing a path model. *Journal of Cardiovascular Nursing, 34*(4), 297–305.

20. Patricia, A. G., & Lisa, L. G., (2014). self-management: A comprehensive approach to management of chronic conditions. *Am. J. Public Health., 104*(8), e25–e31. https://dx.doi.org/10.2105%2FAJPH.2014.302041.

21. Rahmet, G., Imran, H., & Firdevs, A., (2020). COVID-19: Prevention and control measures in the community. *Turk J. Med. Sci., 50*(SI-1), 571–577. doi: 10.3906/sag-2004-146.
22. Ali, F., Samuli, L., & Najmul, I. A. K. M., (2020). Impact of online information on self-isolation intention during the COVID-19 pandemic: Cross-sectional study. *J. Med. Internet Res., 22*(5), e19128. doi: 10.2196/19128.
23. Ruiz-Roso, M. B., et al., (2020). COVID-19 lockdown and changes of the dietary pattern and physical activity habits in a cohort of patients with type 2 diabetes mellitus. *Nutrients, 12*(8), 2327. https://dx.doi.org/10.3390%2Fnu12082327.
24. López-Moreno, M., Maria, T. I. L., Marta, M., & Garcés-Rimón, M., (2020). Physical and psychological effects related to food habits and lifestyle changes derived from COVID-19 home confinement in the Spanish population. *Nutrients, 12*(11), 3445. doi: 10.3390/nu12113445.
25. Xinning, G., et al., (2017). Understanding the patterns of health information dissemination on social media during the zika outbreak. *AMIA Annu. Symp. Proc.,* 820–829.
26. Nawal, A. Al. E., & Boshra, A. A., (2020). Crisis and disaster management in the light of the Islamic approach: COVID-19 pandemic crisis as a model (a qualitative study using the grounded theory). *J. Public Aff.,* e2217. doi: 10.1002/pa.2217.
27. Yan, J., (2009). The effects of public's cognitive appraisal of emotions in crises on crisis coping and strategy assessment. *Public Relations Review, 35*(3), 310–313. doi: 0.1016/j.pubrev.02.003.
28. Elham, M., & Alireza, H., (2020). The role of telehealth during COVID-19 outbreak: A systematic review based on current evidence. *BMC Public Health, 20,* 1193.
29. Carol, L. J., (2020). Parasocial interaction, the COVID-19 quarantine, and digital age media. *Human Arenas.*
30. Bagcchi, S., (2020). Stigma during the COVID-19 pandemic. *The Lancet Infectious Diseases, 20*(7), 782. doi: 10.1016/s1473-3099(20)30498-9.
31. Fatih, O., (2007). Social support and resilience to stress from neurobiology to clinical practice. *Psychiatry (Edgmont), 4*(5), 35–40.
32. Sophie, B., Alexandra, D. C., Aric, A. P., & Andrew, S., (2019). Mindfulness on-the-go: Effects of a mindfulness meditation app on work stress and well-being. *J. Occup. Health Psychol., 24*(1), 127–138. doi: 10.1037/ocp0000118.
33. Lee, C. B., Chen, M. S., Powell, M. J., & Chu, C. M. Y., (2015). Self-reported changes in the implementation of hospital-based health promotion in Taiwan. *American Journal of Health Promotion, 29*(3), 200–203. doi: 10.4278/ajhp.120816-arb-397.

CHAPTER 10

Theoretical Framework on the Need of Emotional Balance and Work-Life Balance for Employees in the Epidemic Situation: A Reference to COVID-19

NEELNI GIRI GOSWAMI

Assistant Professor, University School of Business-Commerce,
Chandigarh University, Chandigarh, India,
E-mail: neelnigoswami@gmail.com

ABSTRACT

Something very serious is happening across the globe! A deadly COVID-19 spreads everywhere and creates critical conditions everywhere. In view of the health and safety organizations are forced to allow work from home to their employees. Organizational changes do not always achieve their expected outcomes and may have a negative impact on employee's well-being and emotional balance. In the present study, the researcher tries to explore the employee's perception on forced work-from-home initiatives. Resistance to change is always been part of the system, hence it is very important to know the impact of the initiative on employee's emotional balance and work-life balance. The employees are bound to work in new setup they require some time to adjust because every organization does not provide the option of work from home to their employees. The epidemic is creating disaster across globe in context of health and economy but maybe it has positive side too. It enables employees to spend time with

---

Emotional Intelligence for Leadership Effectiveness: Management Opportunities and Challenges
During Times of Crisis. Mubashir Majid Baba, Chitra Krishnan, & Fatma Nasser Al-Harthy (Eds.)
© 2023 Apple Academic Press, Inc. Co-published with CRC Press (Taylor & Francis)

their families, which they only dreamed off. So, the purpose of the study is to review the employee's perception of forced work from home initiative. The interview method was used to collect the data from employees. The results indicate that (90%, N=100) maximum respondents are enjoying their family time. The study suggested that organizations should think about the work-from-home initiative seriously and allow employees to get benefited whenever they needed. They also try to use IT in more useful ways like converting offices into e-offices, HRIS (human resources information system), etc.

10.1 INTRODUCTION

As COVID-19 spread throughout the globe, the big players in the market are suffering badly. Rapidly spread of the virus also creates trouble for the global investors in the market, which is directly affecting the economies of the individual nation. Unfortunately, the organizations could not stop working for a longer duration, which resulted in finding new ways to complete their targets *"Work from home"* or *"Forced work from home."* Everybody is talking about the impact of the problem across the nation. As COVID-19 affected cases are on the rise in the countries, organizations are bound to take preventive measures to avert the spread of the virus. Many big organizations like Google, Cognizant, Paytm, Accenture, and Wipro have stepped to secure their employees from epidemic spread [12]. Companies are sanitizing, keeping hand sanitizers on the entry and exit gates, banning the social gatherings, distributing the N95 masks, and disinfecting the office premises to prevent the employees. Many IT companies are taking forward steps as their employees are frequently travel cross boundaries and may contact the deadly virus. Financial sector is struggling as interest rates are going down that to borrow money is almost essentially free. Emre Tiftik, director of Research for Global Policy Initiatives at the Institute of International Finance, a Washington-based financial industry trade group quote, "We have been always saying that we are sitting on top of an unexploded bomb, but we don't know what is going to trigger it" [5]. In many countries, the situation is getting worse due to COVID-19, so organizations are doing rapid preparations which never considered enabling their employees to work from home. Looking at the emotional impacts on those who suddenly force to work in this setup until offices will be reopened with a new set of challenges.

Theoretical Framework on the Need of Emotional Balance 143

Many working groups who are focusing more on remote working are joining their hands together behind the hashtag "#remoteagainstCOVID-19virus" encouraging others to share tips and strategies to start remote working sooner. David Prince, workplace wellbeing expert and CEO of Health Assured, discussed that "it is difficult to separate your work and professional life especially when they both exist in the same place but at the same time remote workers are more engaged and loyal towards their work" [14]. The continuous and rapid effects of COVID-19 worldwide are creating unpredictable level of uncertain changes in both the personal and professional life of the employees.

10.2   NEED OF THE STUDY

The COVID-19 has now reached a new level where things are becoming serious and out of control. At this time, public health systems need to act decisively and actively to resist the growth in the new epidemic time. Now the main focus should be on containing and mitigating the diseases itself. Companies are trying to deal with the situation and implementing innovative strategies to secure their employees physically and emotionally. Everybody is talking about the negative impact of the COVID-19 virus on the health of the nation, but it is taking a toll on their work-life balance and emotional balance. Home becomes office now where employees are working 24×7 virtually. Different perceptions and experiences are shared by many articles like while working from home, employees are getting time with their families, cleaning of home and clothes, and so on. Many companies across the globe are asking their employees to work from home and take unpaid leave. These types of disruption to daily work life have been confined to many parts of the countries across the globe. For many employees, the situation is not easy because of sudden change in the schedules where it is very difficult for an employee to be emotionally balanced and stress free. The call of an hour creates impetus for the present study where researcher trying to collect, analyze, and evaluate the different perception employees who are part of forced "work from home" initiative. As the epidemic situation still continues, researcher did not find any such study that has focused on the real and different perceptions of employees on forced work from home or home quarantine experience which proves the novelty for the present study.

10.3 LITERATURE REVIEW

Speaking to IANS (Indo-Asian News Service), Jaipur State Women and Child Development Minister Mamta Bhupesh shared that "Kids are now happy because they are getting food being cooked by me and are terming it as Mummy ka Dhaba. Also, in the absence of domestic help, I am looking after my garden, cleaning my home and vessels and what not. During Navratri time, I offered my regular prayers to Mata Rani so that everything turns normal for everyone. We never had extra time for these prayers so now I am giving my time to prayers. Also, I am back to my favorite hobby – cooking – which I was not getting time for despite being a good cook. My son and daughter-in-law are helping me in all other chores as the domestic help is also off" [9]. An oil company and a media house asked their employees to work from home in London. Similarly, the media group OMG (Omnicom Media Group) also sent their 1,000 employees back to home after a staff member recently traveled through Singapore showing symptoms. The British-pay-television company started screening their visitors and keep away from their regular employees until they are not medically checked. Lufthansa airline stopped the recruitment process and asked employees to take unpaid leaves as they are facing an economic slowdown due to the present situation. Again, one more advertising agency Dentsu asked their employees to work from home. Milan is not a closed city but it is drastically slowdown in the scenario. The insurance giant Generali and Armani have adopted "smart working" policy to secure their employees [8]. Apple CEO Tim Cook said in an interview that "I think of this as the third phase in getting back to normal." Cisco Systems develop and sell online meeting and e-conferencing application experienced drastically outbreak in the number of users. Traffic on several Webex's routes in China has increased nearly 22 times because of the situation people are commuting from their locations. JetBlue Airways help people in rescheduling and canceling their fights for new booking. American Airlines suspended cancellation and rebooking fees of the scheduled flights. Amazon also sent workers on leave when one of the delivery people was found COVID-19 positive in the facility [7]. One of the workers from ETX (embedded technology extended) who is a mom as well stated that as the COVID-19 outbreak continues, it is not easy to find a home and work balance [3].

Theoretical Framework on the Need of Emotional Balance 145

10.4 RESEARCH DESIGN AND METHODOLOGY

10.4.1 RESEARCH DESIGN

The research design is exploratory in nature. The researcher is trying to investigate a problem which is not clearly defined. The study conducted to have a proper understanding of the existing problem. The researcher started with the general idea and will use this research as a mode to identify issues that can be considered for future research.

10.4.2 DATA COLLECTION

The data was collected through both primary and secondary sources. The secondary data was collected through available sources in the form of articles, news, etc. As the scenario still continues, there is a scarcity of available literature. The primary data was collected in the form of observations and interviews from 100 employees and business persons. Due to critical environment, there was no possibility for the researcher to have face-to-face interviews, so the researcher conducts telephonic interviews and circulate open ended question through Facebook messengers, WhatsApp, and different online apps. The respondents for the study are employees working in any sector and experience work from home.

10.4.3 TIME AND LOCATION

At the time of epidemic (COVID-19) and the data collected on PAN (Presence Across Nation) India basis.

10.5 ANALYSIS AND RESULTS

The primary data was collected in two parts. The response rate was 90%. In the first part, the researcher asks respondents three statements referred from Statista Research department 2020 where respondents give their answer is YES or No. In the second part researcher as respondents about how COVID-19 is affecting the employee's emotional balance and work-life balance. Organizations are adopting various strategies like work from home, virtual meetings, shutting down various public places and so on. Hence in a way employees are getting their "Me time or We time." Researcher ask respondents to share their views as an employee how this situation is affecting their emotional balance and work-life balance? Researcher ask them to make suggestions as well. On the basis of information given by Statista Research department 2020, the researcher asked the same question while interviewing the respondents which are presented in the form of values in Figure 10.1.

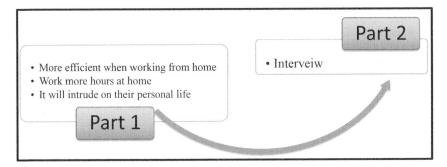

FIGURE 10.1 Data analysis diagrammatic representation.

10.5.1 ANALYSIS FOR PART 1

The data collected through secondary source in Figure 10.1 indicate that 30% of the employees believe that they are more efficient when they are working from home. Employee shared that the reason for being less efficient at home is that at office they work in a proper set up, colleagues are near but now they need to call for every little thing which is quite frustrating. Around 50% agreed that they are working more serious hours at home in comparison to office hours because there is no disturbance in between as time limits are set by them only. Around 37% believe that work from will intrude their life but they are working from home where mentally they are with their family and physically, they are working which create emotional imbalance. Majority agreed they are enjoying best time with their families but in between they attend their office task as well (Table 10.1).

Theoretical Framework on the Need of Emotional Balance 147

TABLE 10.1 Statistics Related to the Efficiency of Employees During Work from Home

Statement	Response (%)	
More efficient when working from home	28	
Work more hours at home	52	
It will intrude on their personal life	37	

Source: Home office and work-life balance of employees in India at the time of COVID-19.

10.5.2 ANALYSIS FOR SECTION 2

Responses of the employees interviewed and observed are reproduced here. The data collected was collected in qualitative form, so it was not possible for the researcher to present all responses. As the responses were repetitive in nature, on the basis of similarity, they were combined and are reproduced in Table 10.2.

10.6 DISCUSSION

From the collected data, it is very much clear that most of the employees are spending quality time with their family where they do not need to keep a watch on time. Otherwise in the routine scenario to have social life for them means office colleagues and peers because they do not get enough time to be with friends, neighbors, and relatives. As in the running time the work encompassing a large amount of time in employee's lives. It is not at all surprising that now employees are seeking for the meaning and purpose behind their professional life. One of the approaches to meaningful work is occupational calling. "Calling refers to the perception of personal meaning, dedication, and involvement tied to one's career." So, employees who find meaningful ties with their work spend long hours in the office premises and make many compromises on their emotional and social parts to live out their calling [16]. Apart from the work-life, the vicious circle of work life balance includes personal, family, and social life. Due to the target deadlines organizations are forcing their employees to work from home in the pandemic situation. It clearly indicates that not

TABLE 10.2 Responses Collected from the Respondents

Profile	Work Experience in Years	Responses
IRPFS, Div. Sec. Commissioner	9	As Indian railways has suspended passengers' trains services and only goods train are functional, workload on railway security staff has reduced a bit. My nature of job is such that complete work from home is not possible. However, as the offices are closed, I am managing the office work from home with some surprise field visits so as to lift the morale of staff and also have better command and control over my team. There is a daily video conference with the honorable Minister of railways which officers take from home in which issues and challenges are discussed and solutions are delivered to field level staff. I am spending more time with family and developing skill sets which help me to perform my duty even better in future.
Assistant Professor	1	Due to COVID-19 my college is closed for the entire month. My personal and professional life is becoming like a new flavor of "Haldirams mixture namkeen"! I get less time with my husband because we both keep on working for the whole day. To complete the syllabus, I am taking online classes and give assignments to complete my syllabus. On the other side, I really enjoy the new experience which gives me an opportunity to learn something new. I am getting familiar with the different uses of the IT which I was ignoring earlier. I support the efforts put by the government in eradicating the COVID-19 from the country. I am not going out in public places and maintaining hygiene standards as well by using sanitizers and masks.
Deputy Manager of Collections with BMW India Financial Services Pvt. Ltd.	13	In my 13 years of work experience, I never have been given an opportunity to work from home. The industries I have been associated with never had such set. Now, due to COVID-19 outbreak companies had to take a decision of putting each of their employees on work from home and within a week's time they found alternatives and IT setups to successfully implement the same. Working from home has pros and cons both. If I talk about pros, I have been able to look after my house and been able to get a lot of chores done by my domestic help which I am usually unable to do due to my busy schedules. I have been able to spend more time with my husband. He is around even when I am working or on office calls. His presence around me gives me happiness. I have been able to look after my health and work in my comfort zone without any unwanted interruption. Apart from personal benefits, there are few good things that have happened. The traffic and pollution are less

TABLE 10.2 *(Continued)*

Profile	Work Experience in Years	Responses
		on roads. I feel the crime rate has also gone down as people are scared to touch others. If I talk about cons, working from home for too long gets monotonous and you need to go to the office because you have people around you to socialize. The screen time has increased even more by being home because we are working from our laptops and phones. When we go out, we have a set schedule for everything. However, when we are at home, we do things at our own convince which is the best part. I must say that every employee should have a flexibility to choose if they want to work from home or office.
Marketing Executive at Pharmaceutical Company	17	I am working as a marketing executive at pharmaceutical company from last 17 years. We deal in direct marketing so work from home is not our cup of tea. Infect when the problem is related with the medical field then it is our busy time as companies are trying to promote their medicines and drugs. Every day I used to visit some hospitals to meet doctors and recommend our medicines. So only doctors are not working anybody who is affiliated with the medical field is working directly or indirectly. I appreciate that the strategies are adopting by the public and private sectors are for employee benefits. They all protective materials like mask and sanitizers. Though I need to work but government is also taking care of those employees who are working in the field.
Executive at Accenture	1	My social life gets affected a lot when I have to work from home because when I am at the office, I meet friends and colleagues. I can interact with them and get the time to chill my mind. But when I am working from home and due to the current situation cannot go out, I feel like a prisoner because during office hours I have to work and remaining time I can out. So basically, I am trapped in a room for all day. Sometimes I feel like I am on an island. Organization and government have taken the right step there should be a lockdown for it otherwise the problem will continue.
IT-based Professional	7	I need to be on the phone for the whole day as per my work schedules due to which I am not able to spend much time with my kids and wife. I remain locked in one room for the whole day as I need peace and isolation to attend the calls but not getting so makes me frustrated and inactive. I really hope that situation outside will get better soon and I can go out to socialize.

TABLE 10.2 *(Continued)*

Profile	Work Experience in Years	Responses
Research Assistant at Indira Gandhi National Center for the Arts	2	I have a fond of reading books. Even when I visit to my friend's place, I carry a book to read in small free time. I joined this organization two years back, and we are working on various projects related to the extinction of various cultures from the society. We are working insane on sights and lots of traveling is also included. Now, as the situation is critical outside, I am really enjoying my own time. I sleep a lot, read as many books I can in a day and my favorite 'Chai (Tea)', what else any literature lover needs.
Entrepreneur	11	I am working as an entrepreneur from last 6 years. Prior to this I worked as a mechanical engineer with one of the good brands. I establish everything on my own which needs most of my time on work. I was giving average time to my family as I was busy with my struggling phase. In the meantime, I also got married and did not get time to even go for a vacation. In view with the present scenario, social distancing is required, so like others we are not commuting out. Trust me if I ignore my work part this is the best phase of my life as I am spending time with my wife and parents all together. We eat together, joke together and even cook together.
Assistant Professor	4	I am working as an assistant professor and have two kids. One is 5-year-old and other is 2-year-old. My husband runs a food business due to which he came late at night and we hardly get time to talk. I always feel like a pendulum and need to take help from my housekeeping staff to take care of my kids as I have my work commitments as well. Due to the pandemic situation outside I am seriously spending quality time with my family which I never had before and always dreamed of.
Housewife	3 months	I am a housewife now as I recently got married and shifted to Germany with my husband. I left my job. We are just three-months-old couple. Due to my Visa process, I shifted late to Germany. Here situations are also critical and companies are asking their employees to work from home. My husband remains on office calls for the whole day but I am satisfied that at least he is near to me. We are having breakfast, lunch, and dinner together. Enjoy long gossips and do the household tasks together makes our bond stronger.

Theoretical Framework on the Need of Emotional Balance 151

TABLE 10.2 (Continued)

Profile	Work Experience in Years	Responses
Script Writer and Event Planner	6	I commute frequently between Delhi and Lucknow as my father keeps unwell. I am home quarantine from last one week due to which I am unable to visit my home town to see my father. My mother and brother are managing everything at home. I enjoy my free time because my profile includes lots of traveling, but same time. I want to meet my family which I miss the most. I am appreciating the step took by the government but everything has dark and the brighter side both. I am a fitness freak but always skip my workout due to my work nature. These days I am completely enjoying my workout and working towards my health.
Marketing Professional	10	I am working as a marketing professional with the company dealing in crushers. I am very busy with my schedule like 25 days in a month I travel. My wife was also working, but after kids she left the job. In a way she alone bought up our kids. I never get time to spend with my family. But now I am spending quality time with my family and dealing with my clients online. I am realizing that what precious I missed in my life while running for my occupational calling. I get enough time with my wife and kids. We plan things together; I love to cook so these days my wife is on complete resting zone.
Deputy Manager at Telecom Company	15	I am working with telecom industry which include handling of large number of labors, so I need to be continuously on phone to take updates about sites. I recently got transferred to another location. and as we recently had our second child, it is not possible for my wife to commute. I kept on commuting every week between the two locations. We are planning to shift by the end of the month and unfortunately states are shut down and people are not permit to commute. I was on my site and somehow managed to come back. Now I am at home taking care of my kids and wife, as we are staying in nuclear set up so my wife was managing all alone. It is impossible to manage work and family together when you are working from home but yes, I agree that companies should give this option when their employees need it.

TABLE 10.2 (Continued)

Profile	Work Experience in Years	Responses
HR Manager	8	I wish organizations showed genuine interest in employees' health and safety rather than paying lip service and making some dramatic show. I do not think employees appreciate organization for this as most of them were pushing commercial interest and asking people to come, until it became mandatory.
AVP, JP Morgan	10	Initially, it was good to know that our company cares for this much that we are given the privilege to work from home. Personally, it was great as I am having a year and quarter old kid at home. I thought I would be able to spend quality time with him. But when I started working from home, there were so many connectivity challenges like I cannot expand my screen at home, network problem. Yes, and one thing I really enjoy is that there is no everyday pressure of formals and reach office on time.

Theoretical Framework on the Need of Emotional Balance

all organizations are concerned about their employee's emotional health and safety rather they are bound to do so under the pressure of governing bodies. Somewhere unknowingly employees are spending quality time with their own selves and family. They are cooking, washing, and cleaning with their families. Though they get less time as they are working from home but they do all the stuff that makes them happy like painting, singing, dancing, relaxing, and so on. In the race of proving better from others, everybody is living like robots and working clockwise. There is a fixed schedule for everything, people are forgetting to do random things in their life and feel threaten in changing their schedules. From family picnic to marriage everything is dependent upon the organization's calendar. People are delaying their outings with family, cancel family plans, delay marriage often just to fulfill organizational commitments. Salaries are going up, but a spark is missing in the workforce. Gradually, the old school of thought of management again entering into the market where the only focus was on target accomplishments. If talking about personal life they are getting time for their hobbies and interest. In the earlier studies of the researcher [6], it was concluded that when employees get time to do things of their choices, they feel emotionally balanced and creative. In the same line, family, and social part also cannot be ignored. Everybody is staying in a nuclear setup where casual relationships are existing these days. There is not a much difference between a home and hostel. With the forced "work from home" concept employees are getting in touch with their families directly and indirectly. They call their friends, have long chats, late night movies with families, playing with siblings and kids, be in a lazy mode for the whole day, no worries about looks, and most important, parents are getting time with their children.

10.7 CONCLUSION

It is a time of reflection for the employees. My life, my choice, my freedom, my independence, my right, and my habits should be focused by all employees. No, means absolutely not. Your choice, habit, freedom can become hindrance in sustenance of world. Simple living, restraints, reasonable restrictions, self-disciple all are necessary for the nourishment of the world. Future of work is so near. Taking this opportunity of *"forced work from home"* employers and employees can analyze the effectiveness of this mode of working. If it works fine then so many communications, traffic jams, travel frustrations, environmental degradation can be avoided. Cognitive dominance [15] hits every employee so hard that they forget everything and became a part of the race blindly. Every employee is working to have a big house and car, nobody cares about basic things in life. As the Maslow hierarchy need theory indicate that the first level in anyone's life is physiological need or basic needs of food, clothing, and shelter. The people of today's generation are taking much time at the basic level of hierarchy which delay many things in their life like buying a house, getting married, children, and much more. To work is an everyday routine for employees but to work from home is a new challenge which everybody is not comfortable with. This new strategy is helping them in making their emotional relationships strong with their partner, children, and elders. They are getting time for their health, hobbies, and can get in touch emotionally with their friends. Employees are having different experiences with work from home but one thing they all demand is that it should be optional for every employee on a regular basis. They must have a leverage to have work from home whenever they want. Organizations do not worry about their work because in the present situation also, employees are working with all dedications, which shows their organizational commitment but goal accomplishment in the absence of emotions is a matter of concern for both employee and employer both.

10.8 SIGNIFICANCE OF THE STUDY

Though the initiative is forced but on the basis of literature review and collected data in the form of observation and interview, it is transparent that work from home should not be forced. It must be provided to the employees as per their requirement because the phase of life for every employee differs. The study will help organizations to realize that employees are part of the organization and their emotional participation matters in framing policies and strategies. In the span of time, the line between work and life is getting lighter which ultimately results in employees who are working without emotions, consistency, and satisfaction. If the scenario continues, then organizations will soon face the problem of employee sustainability. The study will help in underlining the feelings of employees while working from home, ignoring the fact that it was forced.

10.9 SUGGESTIONS

Succession planning is the need of an hour! This time will work as an alarm for everyone including employees, organizations, and government. There is a need for organization now to think about the working patterns they are designing for their employees. Like 360° appraisal there should be 360° participation procedure in the organizations. Every employee wants to work accordingly but only software professional get the leverage to work from home. Now as due to the scenario, all organization are bound to allow their employees to work from home clears many doubts. They should give some time to employees in such situations as it is also difficult for them to adjust to the new working environment. Organizations should allow their employees to opt for the option of working from home when they need it. They should have proper facilities in their premises especially related to health and safety. Time to time every person working in organization should provide training about basic hygiene and make their employees more technically advanced, so in future also if similar situation will arise then organizations will not suffer because its directly related with the economy. If meetings can be organized through e-conferencing then in future also organization will continue with it, which also helps in environment sustainability. Researchers and respondents both believe that this is an opportunity for Indian organizations to redevise their strategy to cater demands of people and strive for paperless working by speeding up various interventions like e-office, HRMS, HRIS, etc. Employees should also try to complete their task within the allotted time frame as future is very unpredictable now.

KEYWORDS

- COVID-19
- emotional balance
- employee perception
- environment sustainability
- epidemic
- global investors
- methodology
- work-life balance

REFERENCES

1. Britt, C. A., (2020). Living to work: The role of occupational calling in response to challenge and hindrance stressors. *Work & Stress.*
2. ET, (2020). *What Big Companies Are Doing to Prepare for Coronavirus Fallout.* Retrieved from: BARRON'S: https://www.barrons.com/articles/what-big-companies-are-doing-to-prepare-for-coronavirus-fallout-51582927415 (accessed on 5 July 2022).
3. Cbs19.tv, (2020). *ETX Mom Talks About Finding Work, Home Life Balance amid COVID-19.* Retrieved from: CBS 19: https://www.cbs19.tv/video/news/etx-mom-talks-about-finding-work-home-lifebalance-amid-covid-19/501-5f630e9a-f215-4ed1-bd2d-80b04fc7d8fe (accessed on 5 July 2022).
4. *Getting Your Workplace Ready for COVID-19.* (2020). Retrieved from: WHO: https://www.who.int/docs/default-source/coronaviruse/getting-workplace-ready-for-covid-19.pdf (accessed on 5 July 2022).
5. Goodman, P. S., (2020). *Coronavirus May Light Fuse on 'Unexploded Bomb' of Corporate Debt.* Retrieved from: Economic Times: https://www.nytimes.com/2020/03/11/business/coronavirus-corporate-debt.html (accessed on 5 July 2022).
6. Goswami, N., & Nigam, S., (2018). Working woman or superwoman: An empirical approach to study the work-life balance among working women in India. *International Journal of Engineering Technology, Management and Applied Sciences, 3.*
7. Heater, B., (2020). *Workers Sent Home After Amazon Warehouse Employee Tests Positive for COVID-19.* Retrieved from: Techcrunch: https://techcrunch.com/2020/03/19/workers-senthome-after-amazon-warehouse-employee-tests-positive-for-covid-19/ (accessed on 5 July 2022).
8. Horowitz, J., (2020). *Coronavirus Stalls Milan, Italy's Economic Engine.* Retrieved from: The New York Times: https://www.nytimes.com/2020/02/24/world/europe/24 coronavirus-milan-italy.html (accessed on 5 July 2022).
9. IANS, (2020). *COVID-19: Rajasthan Woman Politicians Strike Work-Life Balance.* Retrieved from: Twistarticles: https://twistarticle.com/covid-19-rajasthan-woman-politicians-strike-work-life-balance/ (accessed on 5 July 2022).
10. Johnson, J., (2020). *Impact of Home Office on the Work-Life Balance of Marketers During the Coronavirus Pandemic in the United Kingdom (UK) in 2020.* Retrieved from: Statista: https://www.statista.com/statistics/1104762/coronavirus-home-office-and-work-life-balanceof-uk-marketers/ (accessed on 5 July 2022).
11. Karina, N. J. D., (2019). What about me? The impact of employee change agents' person-role fit on their job satisfaction during organizational change. *An International Journal of Work, Health & Organizations.*
12. Manali, (2020). *Shut Offices, Work from Home, Travel Ban: India Inc Takes coronavirus Safety Measures.* Retrieved from: Business Today. https://www.businesstoday.in/current/corporate/coronavirus-india-inc-corporates-safetymeasures-shut-offices-work-from-home-travel-ban/story/397630.html (accessed on 5 July 2022).
13. Martin, R. N. L. S., (2020). *Lead Your Business Through the Coronavirus Crisis.* Retrieved from: Harvard Business Review: https://hbr.org/2020/02/lead-your-business-through-the-coronavirus-crisis (accessed on 5 July 2022).

14. Nair, P., (2020). *Mental Health and COVID-19: How to Stay Safe and Mentally Balanced.* Retrieved from: Real business: https://realbusiness.co.uk/mental-health-covid-19/ (accessed on 5 July 2022).
15. Sharma, S., (2006). *Management in New Age: Western Windows Eastern Doors.* New Age International Publishers.
16. Wilson, A. C., (2018). *Living to Work: The Effects of Occupational Calling on Mental Health at Work.*

CHAPTER 11

Managing Emotions in the Pharmaceutical Sector: How to Minimize Negative Emotions

SOUMENDRA DARBAR,[1] SRIMOYEE SAHA,[2] and SANGITA AGARWAL[3]

[1]*Research and Development Division, Dey's Medical Stores (Mfg.) Ltd., 62, Bondel Road, Kolkata–700019, West Bengal, India, E-mail: dr.soumendradarbar@deysmedical.com*

[2]*Department of Physics, Jadavpur University, 188, Raja S.C. Mallick Road, Kolkata–700032, West Bengal, India*

[3]*Department of Applied Science, RCC Institute of Information Technology, Canal South Road, Beliaghata, Kolkata–700015, West Bengal, India*

ABSTRACT

Stressful situations are all too common in the workplace. It may become harder and harder to manage your emotions under these circumstances. Mainly emotion is a complex feeling state accompanied by physiological arousal and overt behaviors. It varies from man to man, person to person. The consequences of emotional states in the workplace, both behaviors, and attitudes, have substantial significance for individuals, groups, and society. Positive emotions in the workplace help employees obtain favorable outcomes including achievement, job enrichment, and higher quality social context. Negative emotions, such as fear, anger, stress, hostility, sadness, and guilt, decreased work performance, reduced productivity,

---

Emotional Intelligence for Leadership Effectiveness: Management Opportunities and Challenges During Times of Crisis. Mubashir Majid Baba, Chitra Krishnan, & Fatma Nasser Al-Harthy (Eds.)
© 2023 Apple Academic Press, Inc. Co-published with CRC Press (Taylor & Francis)

160   *Emotional Intelligence for Leadership Effectiveness*

and effects on product quality. The pharmaceutical industry plays a crucial role in today's world economies. Emotional labor is likely to be common among most employees across several vocational fields, not just those that entail services to the public. The significance of emotional labor has been acknowledged in a variety of occupations. Today, most organizations especially in the pharmaceutical sectors manage or regulate employees' emotions in order to accomplish their organizational goals. These regulations and requirements have been found to be more prevalent in jobs.

11.1 INTRODUCTION

Stressful situations are all too common in pharmaceutical manufacturing industries. To manage the emotion in this stressful environment now became a great challenge for the higher management staff in the industry. It is really to header control the negative emotion in the pharma industry. Emotion is a complicated feeling state directly connected with psychological and physiological behaviors. It varies from person to person, employee to employee. Studies showed that females are more emotional than males. Emotional states in the working areas connected with attitudes and behaviors have substantial significance for employees, working groups, and surrounding society. Emotion is a god-gifted phenomenon that plays a crucial role in the workplace to build a positive relationship, directly increasing productivity. Positive emotion is always good for the industry, whereas negative emotion declines the industries growth and reduces profit. To increase the skill self-regulation, motivation, and self-awareness are involved which decreased depression and stress [1–4]. Embracing the nuances of human feelings in the workplace can have practical benefits, such as improved collaboration between employees and a happier workplace. In industry, Workman emotion, intelligence, and overall temperament play a significant role in work performance, leadership capability, specialized skills, decision-making character, and team sprits. A favorable work environment increased the industry's turnover. Anger and aggressive behavior towards fellow colleagues and management staff develop dissatisfaction which demotivates the workers. Recognition, rewards, and promotional benefits from the management build positive emotional intelligence (EI), most important for better skill development and increased productivity [5, 6]. People who are calm and quiet, good behavior, adjustable, compromising the character and down to earth are happier in life, good, and healthy relation towards colleagues and Handel difficulties and complications [7].

Study of emotional behaviors and their control in the pharmaceutical industry are a flourishing subject in management today. Fortune and Harvard business review (FHBR), a famous business journal published interesting articles on EI. The article focused on various emotional behaviors that are observed in industrial workplaces and their management [8]. Human emotion and stress are lined with hormones like epinephrine, norepinephrine, androgen, and estrogen. Hypothalamus hypophysial portal axis directly controls these hormone secretions through neural connection [9, 10]. Recent research revealed that, for a superior or substandard, emotions impact workman/employees' work performance, decision-making capacity, creativity, quality of work, commitment leads to hampering industrial growth. So, negative emotions in the workplace should be monitored for obtaining a good quality product. Unfortunately, current pandemic employees are suffering mental and social stress which decline EI as a result industry horribly suffers to obtained good quality products. If this situation persists, industries economic stability is questionable [11–13].

The main aim and objective of this research paper is to study the emotional and psychological behaviors of pharmaceutical workers during the lockdown phase and its effects on industrial growth and development. The study was based on both male and female workers in a leading pharmaceutical industry in Kolkata and find out the probable solution to minimize the negative emotion. The paper also examines both EI and negative emotion are important behavioral factors in the workspace. So, maintain the positive emotional climate (PEC) in the pharmaceutical sector through motivational discussion with the employees is the only tool to suppress negative emotion (Figure 11.1).

11.1.1 EMPLOYEE'S EMOTION AND PANDEMIC SITUATION

People all over the globe suffer from COVID-19 deadly infection which is a massive biological disaster in this century. This pandemic situation developed a major global health crisis that severely declines thinking ability and generates various psychological deformities. Not only this but also its reduced the work speed, productivity, performance, etc. Spreading the virus load makes a new challenge for health professionals and scientists. People affected by this virus showed various psychiatric problems that developed negative emotion, developed depression, and anger.

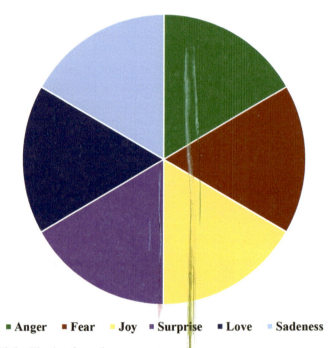

■ Anger ■ Fear ■ Joy ■ Surprise ■ Love ■ Sadness

FIGURE 11.1 Phases of emotion.

The current global health crisis caused by coronavirus not only impacts health but also severely damages the global economy. Health complications not only decline the productivity but also reduce the product quality which decreased the profit. Low income, uncertain job security, family problems, mental disability directly inhibit industrial growth. To overcome this situation is now a new challenge towards the industry people mainly human resources (HRs) personnel [14, 15].

11.1.2 POSITIVE AND NEGATIVE EMOTION

Motivational inspiration in the workplace comes from positive emotions. Positive emotion builds better interpersonal relationships. Good relations

Managing Emotions in the Pharmaceutical Sector 163

between staff and management increased the work power which directly helps for the growth of the company. On the other hand, lots of studies showed that positive emotion makes some personal relationships which improves social interactions.

Negative emotions in the workplace create huge troubles and decreased work performance. When people in the factories are negatively minded to lose productivity. From a management point of view to find out the decline of negative emotion is very essential for a better work environment. Senior managers in the organization have the responsibility to decline negative emotions [16].

11.1.3 STRATEGIES TO MANAGE NEGATIVE EMOTIONS

Based on a recent scientific study, frustration, depression, anger, fear, unhappiness, dislike approach, etc., are the main common negative emotions (CNE) that are frequently observed in the industry. Different situations and emotions are interlinked to each other which directly affects the work performance [17, 18]. The common 10 strategies can be used to control negative emotions. The popular 10 strategies that are extensively used in industries to manage negative emotions (CNE) are (Figure 11.2):

- Compartmentalization (when negative emotions from home affect your work);
- Deep breathing and relaxation techniques;
- The 10 second rule;
- Clarify;
- Blast your anger through exercise;
- Never reply or make a decision when angry;
- Know your triggers;
- Be respectful;
- Apologize for any emotional outburst;
- Never bring your negative emotions home.

11.1.4 EMOTION AND WORK CAPABILITY

Different conditions like stress, emotion, moods, depression, physical activity, mental stability, etc., create an impact on job performance in every industry, especially in pharmaceutical manufacturing industries. Disturb mind decline work performance, decision-making power, creativity, and leadership capacity. The situation decreased the productivity and turnover, which hits the company's stability. In increased work performance motivation training is mandatory in the industry. Understanding people's feelings and behavior change the situation and increased the work capability (Figure 11.3) [19–21].

FIGURE 11.2 Strategies to manage negative emotions.

11.2 EXPERIMENTAL METHODS AND MATERIALS

11.2.1 QUANTITATIVE STUDY

In a quantitative study, all data were collected from a reputed pharmaceutical manufacturing factory based in Kolkata, West Bengal. The study's participants were manufacturing workers and others involved in the different plant operations in the smooth functioning of the industry. The study was carried out during the lockdown period (COVID-19 pandemic) between 25th March and 31st May 2020.

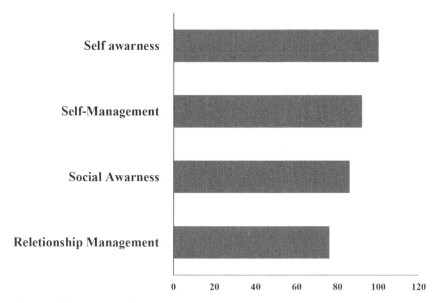

FIGURE 11.3 Stages of emotional management.

11.2.2 DESIGN

Due to the outbreak of COVID-19 severe infection, more or less all the industries of over 210 countries were drastically affected and facing lots of troubles including financial, social, employers, materials, and market More than 3 million people throughout the globe either physically or mentally affected by this infection. To control the spread of COVID-19 infection Government of India declared 21 days Phase-I lockdown from 25th March. After observing the severity of the infection upon the community the Indian government sentenced the lockdown phase till 31st May 2020. As per the government, the circular transport facility was totally suspended during the lockdown period except for emergency and essential service vehicles like medicine, food, water, dairy, fire, police, and other emergency services. All places of worship, educational institutions (schools, colleges, and universities), cinema, and theater halls, shopping malls, and general offices were completely closed during this period.

The study was conducted on male and female pharmaceutical manufacturing industry workers during this lockdown period. Data were collected from 441 workers based on the questionnaire (Table 11.1).

166   *Emotional Intelligence for Leadership Effectiveness*

TABLE 11.1   Questioner for Pharmaceutical Manufacturing Workers

SL. No.	Questioner
1.	Are you frustrated during this pandemic COVID-19 unexpected situation?
2.	Are you worry about this situation and feeling insecure?
3.	Are you angry for this pandemic COVID-19 unexpected situation?
4.	Are you feeling down?
5.	Are you feeling dislike attitude?
6.	Are you anxious?
7.	Are you depressed in this situation?
8.	Are you stressed?

11.2.3 PARTICIPANTS

All the study participants were working in the pharmaceutical company and attending to their work on alternate days in a week from 9 am to 5 pm (day shift duty). The verbal consent of the participants was taken and then only the questionnaire was given to them. The total sample size was 441, which consisted of 387 males and 54 females. The respondents specified their level of agreement to a statement, typically in three points: (i) disagree; (ii) neither agree nor disagree; and (iii) agree. The respondents who participated in the study were debriefed and explained before their responses.

11.3   RESULTS AND DISCUSSION

11.3.1 STUDY OF STRESS PARAMETERS DURING LOCKDOWN PHASE

Out of 441 respondents, the majority of respondents (87.76%) were males. From the responses obtained, it is ascertained that both males and females were frustrated with a response of 83.72% and 85.2%, respectively. The response to the statement on insecurity or worry is almost equal as both males (91.73%) and females (92.6%) were worried during the lockdown phase. The response to the parameter anger was quite different, the response from males (74.94%) > females (40.7%). The feeling "down" syndrome was more in males (83.20%) compared to females (64.8%). The feeling of dislike was almost same in males (77.8%) and females (77.8%). Anxiety was more in males (93.54%) compared to females (87%). Depression was

Managing Emotions in the Pharmaceutical Sector 167

also slightly more in males (91.73%) compared to females (88.88). The males (95.61%) were more stressed out compared to females (92.6%) (Tables 11.2 and 11.3; Figures 11.4 and 11.5).

TABLE 11.2 Study of Emotional and Stress Parameters During COVID-19 Pandemic Lockdown Period (25th March 2020 to 31st May 2020) Upon Male Pharmaceutical Manufacturing Workers

Parameters	Agree	Percentage (%)	Disagree	Percentage (%)	Neither Agree nor Disagree	Percentage (%)
Frustration	324	83.72	57	15	6	1.5
Worry or insecurity	355	91.73	28	7.23	4	1.04
Anger	290	74.94	89	22.99	8	2.08
Feeling "down"	322	83.20	60	15.5	5	1.3
Dislike	301	77.78	80	20.67	6	1.56
Anxiety	362	93.54	18	4.65	7	1.81
Depression	355	91.73	25	6.45	7	1.82
Stress	370	95.61	17	4.4	0	0

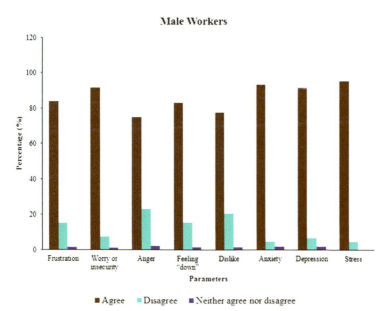

FIGURE 11.4 Percentile representation of different parameters in male workers (sample size 387 males).

TABLE 11.3 Study of Emotional and Stress Parameters During COVID-19 Pandemic Lockdown Period (25th March 2020 to 31st May 2020) Upon Female Pharmaceutical Manufacturing Workers

Parameters	Agree	Percentage (%)	Disagree	Percentage (%)	Neither Agree nor Disagree	Percentage (%)
Frustration	40.0	85.2	7.0	13.0	7.0	13.0
Worry or insecurity	50.0	92.6	3.0	5.6	1.0	1.9
Anger	22.0	40.7	30.0	55.6	2.0	3.7
Feeling "down"	35.0	64.8	14	25.92	5.0	9.3
Dislike	42.0	77.8	6.0	11.1	6.0	11.1
Anxiety	47.0	87.0	3.0	5.6	4.0	7.4
Depression	48.0	88.88	4	7.41	2	3.7
Stress	50.0	92.6	3.0	5.6	1.0	1.9

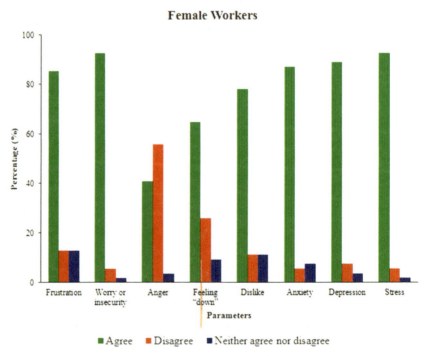

FIGURE 11.5 Percentile representation of different parameters in female workers (sample size 54 females).

11.4 DISCUSSION

The factory has more male workers than females so the percentage of male respondents are more in study and it was not intentional, we did not have equitable distribution of respondents based on gender [22, 23]. The frustration was slightly higher in females compared to males because of the lockdown. They were burdened with household work as well work from home also took its toll. Not only that the children studying at home with schools and colleges closed also impacted the result. The females were more insecure because during lockdown, many people lost their jobs [24–26], they were always worried about jobs security as well as earning the capability of their family members, mainly spouse, children, and others allied members. The lockdown was done in phases so after first phase people thought this lockdown would be over and there was a hope that normalcy would resume but it got extended increasing the uncertainties and worries. The females are less angry than males in this study because they could understand that being angry will not resolve any problem, they instead felt more helpless. The males had more of the feeling "down" response compared to females as few of them rediscovered their hobbies and started baking, making handicrafts, and pursued other activities from home in their off times. This involvement made them fare better than males. The response to dislike was the same in both males and females, as no one wanted a changed work, home front and environment. The response to anxiety, depression, and stress is more in males in comparison to females because most of females have other involvement in household work, activities related to children and others besides working in factories. The concept of being social had taken a backstage and being isolated from friends, family, and peers led to these feelings.

11.5 CONCLUSION

At workplace generally emotion affects attitudes and behaviors. Positive emotion (PM) improves the quality whereas negative emotion decreased the productivity. So, in pharma industries, emotion directly affect the productivity that's why control emotions are the prime factor for the betterment of the organization. If we scientifically and psychologically manage people's emotions, to overcome the challenges and hurdles in the shop floor became easier. Changing the higher boss, departmental rotation, transfer the woman, developed the team working attitude or avoid conflict of interest with a co-worker or customer makes the better work environment during this pandemic situation which increased the productivity yield. In the field study, female workers facing more insecure, and frustration in comparison to males. On the other hand, male workers are more depressed, anxious, and feeling stress in comparison to females. During the unexpected COVID-19 pandemic outbreak, both male and female pharmaceutical manufacturing workers were emotionally stressed and depressed, and worried about insecurity.

KEYWORDS

- **coronavirus**
- **COVID-19**
- **distress**
- **emotion**
- **global health**
- **psychological factors**
- **stress**

REFERENCES

1. Adalja, A. A., Toner, E., & Inglesby, T. V., (2020). Priorities for the US health community responding to COVID-19. *JAMA, 323*(14), 1343, 1344.
2. Chen, Z., Allen, T. D., & Hou, L., (2020). Mindfulness, empathetic concern, and work-family outcomes: A dyadic analysis. *Journal of Vocational Behavior, 119*, 103402.
3. Côté, S., DeCelles, K. A., McCarthy, J. M., Van, K. G. A., & Hideg, I., (2011). The Jekyll and Hyde of emotional intelligence: Emotion-regulation knowledge facilitates both prosocial and interpersonally deviant behavior. *Psychological Science, 22*(8), 1073–1080.
4. Diefendorff, J. M., Richard, E. M., & Yang, J., (2008). Linking emotion regulation strategies to affective events and negative emotions at work. *Journal of Vocational Behavior, 73*(3), 498–508.
5. Brooks, S. K., Webster, R. K., Smith, L. E., Woodland, L., Wessely, S., Greenberg, N., & Rubin, G. J., (2020). The psychological impact of quarantine and how to reduce it: Rapid review of the evidence. *The Lancet, 395*(10227), 912–920.

6. World Health Organization, (2020). *Mental Health and Psychosocial Considerations During the COVID-19 Outbreak.* World Health Organization.
7. Giardini, A., & Frese, M., (2006). Reducing the negative effects of emotion work in service occupations: Emotional competence as a psychological resource. *Journal of Occupational Health Psychology, 11*(1), 63.
8. Wang, C., Pan, R., Wan, X., Tan, Y., Xu, L., Ho, C. S., & Ho, R. C., (2020). Immediate psychological responses and associated factors during the initial stage of the 2019 coronavirus disease (COVID-19) epidemic among the general population in China. *International Journal of Environmental Research and Public Health, 17*(5), 1729.
9. Greenberg, N., Docherty, M., Gnanapragasam, S., & Wessely, S., (2020). Managing mental health challenges faced by healthcare workers during COVID-19 pandemic. *BMJ, 26,* 368.
10. Eisenberg, N., (2000). Emotion, regulation, and moral development. *Annual Review of Psychology, 51*(1), 665–697.
11. Folkman, S., Lazarus, R. S., Dunkel-Scetter, C., DeLongis, A., & Gruen, R. J., (1986). Dynamics of a stressful cognitive appraisal, coping, and encounter outcomes. *Journal of Personality and Social Psychology, 50*(5), 992–1003.
12. Pedrosa, A. L., Bitencourt, L., Fróes, A. C., Cazumbá, M. L., Campos, R. G., De Brito, S. B., & Silva, A. C., (2020). Emotional, behavioral, and psychological impact of the COVID-19 pandemic. *Frontiers in Psychology,* 11.
13. Nachega, J. B., Grimwood, A., Mahomed, H., Fatti, G., Preiser, W., Kallay, O., Mbala, P. K., et al., (2021). From easing lockdowns to scaling up community-based coronavirus disease 2019 screening, testing, and contact tracing in Africa—Shared approaches, innovations, and challenges to minimize morbidity and mortality. *Clinical Infectious Diseases, 72*(2), 327–331.
14. Monaco, A., Manzia, T. M., Angelico, R., Iaria, G., Gazia, C., Al Alawi, Y., Fourtounas, K., et al., (2020). Awareness and impact of non-pharmaceutical interventions during coronavirus disease 2019 pandemic in renal transplant recipients. In: *Transplantation* Proceedings (Vol. 52, No. 9, pp. 2607–2613).
15. Sinclair, R. R., Allen, T., Barber, L., Bergman, M., Britt, T., Butler, A., Ford, M., et al., (2020). Occupational health science in the time of COVID-19: Now more than ever. *Occupational Health Science, 4*(1), 1–22.
16. Canady, V. A., (2020). COVID-19 outbreak represents a new way of mental health service delivery. *Mental Health Weekly, 30*(12), 1–4.
17. Lebowitz, M. S., & Dovidio, J. F., (2015). Implications of emotion regulation strategies for empathic concern, social attitudes, and helping behavior. *Emotion, 15*(2), 187–194.
18. Mukosolu, O., Ibrahim, F., Rampal, L., & Ibrahim, N., (2015). Prevalence of job stress and its associated factors among Universiti Putra Malaysia staff. *Malaysian Journal of Medicine and Health Sciences, 11*(1), 27–38.
19. Reddy, G. L., & Poornima, R., (2012). Occupational stress and professional burnout of university teachers in South India. *International Journal of Educational Planning & Administration, 2*(2), 109–124.
20. Teixeira, C. A. B., Pereira, S. S., Cardoso, L., Seleghim, M. R., Reis, L. N., & Gherardi-Donato, E. C. S. G., (2015). Occupational stress among nursing technicians and assistants: Coping focused on the problem. *Investigacion y Educacionen Enfermeria, 33*(1), 28–34.

21. Teixeira, C. A. B., Reisdorfer, E., & Gherardi-Donato, E. C. S., (2014). Estresse ocupacional e coping: Reflexãoacerca dos conceitos e a prática de enfermagemhospitalar. *Rev. Enferm UFPE Online, 8*(1), 2528–2532.
22. Shammi, M., Bodrud-Doza, M., Islam, A. R., & Rahman, M. M., (2020). Strategic assessment of COVID-19 pandemic in Bangladesh: Comparative lockdown scenario analysis, public perception, and management for sustainability. *Environment, Development and Sustainability, 18*, 1–44.
23. Godinic, D., Obrenovic, B., & Khudaykulov, A., (2020). Effects of economic uncertainty on mental health in the COVID-19 pandemic context: Social identity disturbance, job uncertainty and psychological well-being model. *Int. J. Innov. Econ. Dev., 6*, 61–74.
24. Shammi, M., Bodrud-Doza, M., Islam, A. R., & Rahman, M. M., (2020). COVID-19 pandemic, socioeconomic crisis and human stress in resource-limited settings: A case from Bangladesh. *Heliyon, 6*(5),e04063.
25. Ahmad, A., Rahman, I., & Agarwal, M., (2020). *Factors Influencing Mental Health During COVID-19 Outbreak: An Exploratory Survey Among Indian Population.* MedRxiv.
26. Hamadani, J. D., Hasan, M. I., Baldi, A. J., Hossain, S. J., Shiraji, S., Bhuiyan, M. S. A., Mehrin, S. F., et al., (2020). Immediate impact of stay-at-home orders to control COVID-19 transmission on socioeconomic conditions, food insecurity, mental health, and intimate partner violence in Bangladeshi women and their families: An interrupted time series. *The Lancet Global Health, 8*(11), e1380–e1389.

CHAPTER 12

Emotions, Perceptions, and the Organizational Dynamism Required in Tourism During a Crisis

ZEPPHORA LYNGDOH

Department of Tourism and Hotel Management, NEHU, Shillong–793022, Meghalaya, India, E-mail: zeyadkhar@gmail.com

ABSTRACT

Tourism is an emotional process. However, little is known about the complex nature and dynamic relationship between tourist experiences and their perceptions and emotions about the destination and local communities. The tourism industry has been hailed as the 'fun' industry and is practiced for its benefits and income generation. Tourists choose to spend discretionary disposable income on holidays and travel essentially for the anticipated pleasure that they will obtain. Hence, based on this proposition, tourism is significantly focused on emotions and perceptions. The theorization of emotions and perceptions has received much attention in the contemporary tourism literature and among destination marketer. Emotions, perceptions and episodes of intense feelings associated with a specific situation or event play a key role in understanding tourist behavior. There has been a lot of study pertaining and focusing on the positive emotional experiences associated with festivals, shopping, theme parks, holidays, heritage sites and adventure tourism, among others and the links between emotional responses and behavioral outcomes such as customer/tourist satisfaction and loyalty. Tourist destinations around the world try to focus on and emphasize

---

Emotional Intelligence for Leadership Effectiveness: Management Opportunities and Challenges During Times of Crisis. Mubashir Majid Baba, Chitra Krishnan, & Fatma Nasser Al-Harthy (Eds.)
© 2023 Apple Academic Press, Inc. Co-published with CRC Press (Taylor & Francis)

on the positive emotional connections. For example, countries highlight the surprising component of the tourist experiences in their branding strategies. Some notable successful country campaigns include, "Amazing Thailand" and "Incredible India." These have been built to associate a sense of positive and delight with tourism visits. In marketing and tourism context, perception is considered as a major influential predictor in directing decision making and consumer behavior. Each individual selects, organizes and interprets received information in a unique way. This image depends on both a specific stimuli which are related to the environment and the individual's own characteristics and situations. Destination perception accumulates from destination attributes, both physical and mythical. Tourists' pre-experience a particular destination through various sources about the destination attributes. Destination attributes are commonly used in empirical research to measure tourists' perception of a destination. Tourist destinations consist of a number of attributes that differentiate them from each other. These are listed as accessibility, amenities, accommodation, attractions, and activities. There are also psychological aspects to the tourist experience. Tourists take into consideration most or all of those attributes when making their decision to visit or revisit a particular destination. Potential travelers make comparisons of the attributes of different destinations before they make their choice and decide on the destination that offers those attributes that they deem important. At the same time, different market segments place different levels of importance on different attributes, resulting in different destination choices. Most of the early studies of the effects of tourism focused upon the economic aspects of tourism. However, during the past decade increasing attention has been given by researchers to the social effects of tourism. These social effects are broad ranging and refer to ways in which tourism contributes to changes in value systems, individual behavior, family relationships, collective lifestyles, safety levels, moral conduct, creative expressions, traditional ceremonies, and community organizations. The principal reason for giving attention to the effects of tourism is inevitability that tourism development will induce some impacts. Information about tourism impacts is an important ingredient which needs to be considered in tourism planning. One aspect of social impact research which has been investigated and which provides valuable information for future planning is resident/host perceptions of tourism. The importance of innovation was long underestimated in service activities and the tourism sector. In contrast to the radical innovations vital to growth in manufacturing sectors, innovations in services and tourism were secondary and capital-scarce, and for this reason they were excluded from the scope of government interest and action. It is interesting to note that the discourse changed with the emergence of new information and communication technologies (NICT) and with the COVID-19 pandemic in the year 2020, which have been clearly influential in the realm of tourism. The dissemination of new modes of production and the resulting organizational shock waves, along with the marketing adjustments this has entailed, have been the subject of much research.

12.1 INTRODUCTION

Everyone has heard the word "emotions" in the course of their daily lives. This is transitive because they express a major portion of human attitude, individuality, and decorum. The exchange of emotions has a major impact on the organizations' conduct as well. Primarily, the term 'emotion' delegates "a state of consciousness having to do with the arousal of feelings (Webster's New World Dictionary)." Emotions imply superiority over other mental states like perception, desire/choice, and knowledge of bodily sensitiveness. Emotions are subject to interpretation and often influenced by mood, opinions, psyche, temper, disposal, and penchant. The term 'emotion' originates from the French word Emouvoir which is underpinned on the Latin word *emovere*, where e (alternate of ex) means 'out.' These comprise of either:

- A psychological process that stems instinctively and gradually rather than mediating via deliberate attempts and is often escorted with corporeal and physical changes; including sentiments like: *the emotions of happiness, pain, respect, fondness, and abhor*.
- Being in a *distressed* state and showing signs of *restlessness*: he spoke in a feverish voice or fidgeting that showed intense agitation.
- The state of being aware of and responsive to one's surroundings and involving sense and sentiment; *sensibility*.

A person's behavior is dictated by their perception of their environment, not by what it actually is, but rather by what they believe it to be. People systematize and examine their sensatory ideas in order to have a meaningful impact on the world around them and the environment where they live. As a result, humans have contradicting views of living

176   *Emotional Intelligence for Leadership Effectiveness*

and non-living objects. A person will not make the same inferences about humans as they would about inanimately objects. Everyone has their own individuality and their own set of personal values, beliefs, dogmas, or intentions. Determination and the ability to understand another person's conduct are therefore influenced by these variables. We see the world around us through our senses, which requires us to receive and respond to external cues in a timely manner. By acquiring knowledge about the traits and components of our environment, which are necessary for our survival and which also allow us to function within our environment, we are able to behave within the context of our environment, which is facilitated by the perceptual process. Every organization has a principal commitment to capitalize its resources, both human and material, and employ them to the fullest extent, as has each individual. As a company, businesses have certain responsibilities to society. The goal is to improve the level of living and quality of life for the communities and to sustain profitability for the organization as a whole. It is judged on the economic performance and what it can do for the community, of which it is a member already.

12.1.1   TOURISM AS A BUSINESS ENTITY

Tourism frequently initiates an array of manifold advantages to local communities, improved facilities (electricity, water, utilities, and information technology (IT)), ingress, assistance (banks, communications, travel) and new financing, that all serve to intensify and enrich the lifestyles of the commune [1]. The novel coronavirus (COVID-19) is arduous to the whole world. The tourism sector has been devastated and UNWTO predicted that millions of jobs are also at risk. With limited medical capacity to treat the virus, vulnerable populations in all countries need the vaccination as their highest priority. Unprecedented universal travel limitations and stay-at-home orders have brought about the most serious disturbances to the worldwide economy. Local travel restrictions directly affect economies nationally and internationally, including the tourism mechanisms, i.e., international tourism, domestic travel, tour operators, and segments as diverse as flights, sea trips, public transit, lodging, motels, eatery/diner, conferences, fests, meets or athletic events.

The role of the tourism sector to providing livelihoods and boosting economies is at stake. Owing to the circumscriptions levied on March 2020, whilst the virus started transmitting quickly throughout the world, global travel was discontinued in April and May resulting in the drop in the global tourism business. In the future, the global tourism industry is likely to be affected by many factors, including economic, social, geopolitical, technological, cultural, and environmental. Tourism is a network for cultural heterogeneity, which enables sharing different ideas, traditions, and knowledge between residents (community) and tourists (visitors). Travel is a means to "discover those things that are unknown or forgotten within ourselves." Targeting tourists may involve a mishmash of response driven approaches across a range of markets and localities. Hence, Community support follows progress that serves the community and provides perpetual profits. Tourism development boosts the conservation and transmittal of intellectuality and heritage, preservation, and viable regulation of natural reserves, the safeguard of local traditions, and the restoration of aboriginal customs, artifice, skills, and so on [2]. The intent of the paper is to perceive allegiance towards a destination theoretically by using tourists' approach, destination impression from the residents' purview and tourist satisfaction.

The influences of tourism on every community are complicated and intricate in subject. Tourism effects are vital to different societies, consorts, and people controlled by their Morales, principles, and the type of reserves accessible for advancement in the tourism industry. This study analyzes elements, factors, and traits affecting the destination form and probes into the tourist gratification and rationales of destination fealty. This chapter strives to evaluate current perceptions on tourist contentment and piety to the tourism destination, i.e., Meghalaya after the spread of the novel coronavirus. Tourist Acumen compositions have been controlled by variants like cultural attractiveness, destination reasonableness, safe environment, natural scenery, gaiety events, and organizational structure. Destination imagery compositions have been controlled by variants like primary facilities, traditional attractions, innate attractions, destination cleanness, friendly locals, preservation, maintenance price, and accessibility. The satisfaction composition has been controlled by factors like accommodation, food, transportation, entertainment services, and marketing. The destination loyalty composition is controlled by word of mouth and aims to go back again to tourist destinations [3]. Therefore, we can assess that tourist acumen, destination imagery and satisfaction directly control destination loyalty. Hence, tourism helps reduces privation, fiscal, and affable debarment, and also offers shifting to metro cities [5]. Regardless of the

178   *Emotional Intelligence for Leadership Effectiveness*

profitable aspects of tourism employment and local tourism businesses, it must be remembered that job creation focusing on tourism is capricious, even in developed economies, with notable levels of unpredictability in total sales and temporary jobs [6].

In this study, we ask whether global leaders should have anticipated a pandemic like COVID-19, given that a range of health and economic organizations and institutions have been warning about the increased potential of a devastating worldwide pandemic and if they could recover from the loss that the virus has envisaged in many lives. In the COVID-19 era and beyond, travel, and tourism will be marked by uncertainty and relative tourist concern. Globally, reopening, and resurrection plans will have to be customized to match the specific environment of normal tourism again, given the varying recovery paths.

12.2 METHODS

Data collection was the first and most crucial stage of a survey, as it involves gathering data. As a result, the method of data collection depends on numerous factors such as the purpose of the study, its extent, type, and the availability of resources. For this paper, *a sample of 50 tourists and 50 residents has been taken to study their emotions, perceptions, and tourism development through tourism after the COVID-19 crisis.* The schedule of questions was designed structurally in conformity with the objectives. The scheduled questions comprises of two parts, namely, Part A(Tourist perception) and Part B(Resident perception), which was further divided into distinct sections as per the objectives and variables and to acquire maximum variables based on their individual perception and on Meghalaya as a whole. The subject matter of perception is assessed through a scale of "one," "two," "three," "four," and "five." They stand for "bad," "acceptable," "good," "excellent," and "outstanding." The data was collected from primary and secondary sources.

12.2.1 TOURIST AND RESIDENT PROFILE

As a platform for results and subsequent discussion, Table 12.1 presents the tourist and resident profile with the objective of assessing the homogeneity/difference between the two groups. The profile includes variables

Emotions, Perceptions, and the Organizational Dynamism 179

such as gender, age, literacy level and marital status. In case of gender, the majority of the tourists are male; owing to the fact that female tourists are lesser in number.

TABLE 12.1 Tourist and Resident Profile

Variable	Category	Tourist	Resident	Total
Gender	Male	35	25	60
	Female	15	25	40
	Total	**50**	**50**	**100**
Age	<26	30	20	50
	26–35	17	26	43
	36–45	3	4	7
	46–55	0	0	0
	>55	0	0	0
	Total	**50**	**50**	**100**
Literacy level	Illiterate	0	2	2
	<10th pass	0	0	0
	10th pass	4	7	11
	12th pass	18	15	33
	Graduation and above	28	26	54
	Total	**50**	**50**	**100**
Marital status	Single	31	28	59
	Married	17	20	37
	Separated	2	2	4
	Divorcee	0	0	0
	Widower	0	0	0
	Total	**50**	**50**	**100**

12.2.2 TOURIST AND RESIDENT PERCEPTION OF MEGHALAYA

12.2.2.1 TOURIST PERCEPTION

Table 12.2 portrays the tourist perception. It indicates that most tourists perceive food, transportation, safety, and accommodation as an important aspect while traveling to another place. These variables denote the key factors through which the tourist decides whether the destination is

180   *Emotional Intelligence for Leadership Effectiveness*

amicable or not and if he plans to visit again. While most tourists have found Meghalaya to be of excellent leverage; some tourists found the place to be average.

TABLE 12.2 Tourist Perception

Variable	Score					Total
	1	2	3	4	5	
TF.Q	0	0	12	31	7	50
TF.P	0	0	13	28	9	50
TF.S	0	0	11	30	9	50
TT.Q	0	0	12	26	12	50
TT.P	0	0	12	25	13	50
TT.S	0	0	12	22	16	50
TA.Q	0	0	6	27	17	50
TA.P	0	0	11	20	19	50
TA.S	0	1	8	30	11	50

12.2.2.2 RESIDENT PERCEPTION

In comparison with Table 12.3, where the tourists view an amicable environment and quality in food, transportation, and accommodation. Table 12.3 highlights that the residents are happy to welcome tourists into the destination, and most of them have a fair knowledge about tourism and its prime benefits. But because of the pandemic, they tend to become fearful of the tourists.

12.2.3 CORRELATION MATRIX OF TOURIST AND RESIDENT PERCEPTION

The correlation matrix (Table 12.4) does not show a clear relationship between the perception of tourist and resident. Of the correlation coefficients, only a handful (depicted in bold) are statistically significant. Interestingly, there also exist a negative relationship between tourist and resident perception between some variables. These results portrays that there is a degree of difference on the perception of tourist and resident on Meghalaya as a tourism destination during a crisis. This highlights the

Emotions, Perceptions, and the Organizational Dynamism 181

need and demand for concerted efforts for the development of tourism through the involvement of all the stakeholders.

TABLE 12.3 Resident Perception

Variable	Score					Total
	1	2	3	4	5	
RF.Q	0	1	26	17	6	50
RF.P	0	0	15	29	6	50
RF.S	0	0	15	28	7	50
RT.Q	0	0	11	29	10	50
RT.P	0	0	16	23	11	50
RT.S	0	0	16	27	7	50
RA.Q	0	1	13	26	10	50
RA.P	0	0	13	21	16	50
RA.S	0	0	16	20	14	50

TABLE 12.4 Correlation Matrix

	RF.Q	RF.P	RF.S	RT.Q	RT.P	RT.S	RA.Q	RA.P	RA.S
TF.Q	0.127	0.058	0.010	0.096	0.113	0.206	0.248	0.143	0.204
TF.P	0.010	0.014	0.017	0.090	0.108	0.152	0.108	0.010	−0.006
TF.S	0.180	0.134	0.083	0.047	0.209	0.225	0.209	0.088	−0.044
TT.Q	−0.199	−0.185	−0.135	−0.045	0.040	0.132	0.159	−0.076	0.075
TT.P	0.173	**0.281***	0.271	0.219	0.276	**0.397****	**0.393****	0.110	0.111
TT.S	0.139	0.161	0.194	0.252	0.273	**0.317***	0.273	**0.310***	0.249
TA.Q	0.079	0.049	−0.012	−0.038	−0.039	0.094	0.218	0.014	0.058
TA.P	0.128	0.019	0.011	−0.034	−0.116	−0.023	0.029	−0.121	0.011
TA.S	0.262	0.198	0.236	0.229	**0.288***	**0.414****	**0.450****	0.231	0.116

**significant at 1%; *significant at 5%.

12.2.4 TOURISM: A BOON OR MISFORTUNE

Many people consider the development of tourism to be a vital part of local economies. Development and promotion of tourism has been shown to create new jobs, increase tax receipts and foreign exchange benefits and improve community infrastructure that attracts other companies [7].

Recently, tourism has been seen as a positive economic move, especially in less developed countries (LDCs) [8]. Tourist literature has been increasingly using the term "tourism impact." Local citizens' impressions of the impact of tourism development in their town have been studied extensively in recent years, and this remains a crucial topic. Tourism development can have both positive and bad effects at the local level, according to a growing body of studies [7, 8]. Most often, travel, and tourism are justified by financial rewards, but they are also criticized for harming social and/or cultural values or damaging the environment and its resources [9]. In addition, the economic benefits of tourism expansion are now being assessed against the potential for social disruption, which is growing in importance [8, 9]. Tourism expansion may alter inhabitants' interactions with one other and with their community, according to a new study [10, 11]. A majority of tourism planners and policymakers believe that community views of the impacts of tourism are likely to be an essential consideration in the establishment of present and future tourism programming [12]. Uncertainty (risk) in the process must be taken into account, as well as innovation's role in creating value that is ultimately appraised by consumers [13].

12.2.5 STRESS AND CRISIS MANAGEMENT

Events that generate mental or bodily tension are defined as stress. There is a limit to the amount of stress a person can handle, which can produce mental and bodily imbalance. Brain signals release of stress hormones during handling of stressful situations as a result, the blood sugar levels rise, the heartbeat speeds up, and blood pressure increases. In preparation for action, the muscles tense and tighten. To deal with the circumstances at hand, blood supply is diverted away from the gut and sent instead to the extremities and brain.

12.2.6 GENERAL ADAPTATION SYNDROME (GAS)

A researcher came up with a theory on stress and described systematically that the body reacts to difficult life events like it reacts to disease. This is vital because stress can cause chronic illness, mental fatigue, irritability, and insomnia [14]. He portrayed general adaptation syndrome (GAS), as a flexible approach that takes place in three phases: (i) an alarm stage; (ii) the stage of resistance; and (iii) exhaustion stage (Figure 12.1).

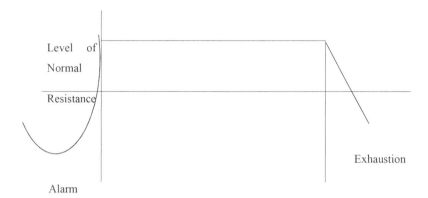

Resistance
FIGURE 12.1 GAS.

1. **Alarm Stage:** According to this model, the general adaptation syndrome begins when a person experiences an irritant and arrives into the alarm stage (under stress). He reflects how to handle the situation and feels apprehension and even nervousness. At this stage, the person tries to cope with uncertainty.
2. **Resistance Stage:** After the initial stage, the person assumes he can cope with the irritant; he/she begins to feel better and the body begins to repair itself. Now the person thinks about how to respond. In this stage, the person focuses his/her vivacity and uses it to withstand the irritants' adverse effects. The person deals with the problem, envoys the challenge, or adapts to the change.
3. **Exhaustion Stage:** Many irritants are temporary–the person tends to solve the issue or the complication ends by itself. In this stage, the traits of the alarm stage reoccur and the person makes use of his/her pliant energy in due course of time.

12.2.7 CRISIS MANAGEMENT

Crisis management is a systematic attempt at preventing or managing organizational crises [15]. In a crisis, an organization and its stakeholders

184   *Emotional Intelligence for Leadership Effectiveness*

are threatened by a large, unpredictable occurrence that threatens to harm the organization.

A threat to the organization, a surprise element, and a limited decision time are all common to most definitions of a crisis [16].

There are four types of organizational crises:

1. **Sudden Crisis:** Like natural catastrophes, explosions, fire outbreak, workplace vehemence, etc.
2. **Smoldering Crisis:** Issues that start out as trivial and could be minimized or forestalled if someone was attentive or acknowledged the possibilities for trouble.
3. **Bizarre Crisis:** Such extreme events which are almost unbelievable.
4. **Perceptual Crisis:** Such as a persistent problem and long-term problem [17].

Notwithstanding the size of an organization influenced, the main advantages of crisis management would comprise of:

- Potential to evaluate the circumstances internally and from the exterior of the organization as the shareholders might get a chance to appraise it.
- Approaches to involve activities to suppress the probable or recognized damage escalation.
- A more constructive method to quickly activate part(s) of business progression control.
- Better managerial elasticity for all the shareholders.
- Assent with organizational and moral requirements. For example, corporate social responsibility.
- Better governance of severe incidents or those that could become hazardous.
- Upgrade staff realization of their roles in the organization and conjectures.
- Amplified capability, credence, and self-esteem.
- Improved risk management, so that evident risks are recognized, reduced (where possible) and intensify business progression management—as planned for.

Example of the COVID-19 as a crisis – the COVID-19 pandemic represents a global health catastrophe of our time and a prominent challenge for everyone. It is also an unmatched socio-economic crisis. The world is facing a global danger, with social and economic emergencies during this pandemic, of which travel and tourism are among the most afflicted sectors with travel restrictions in almost all the countries of the world. The World Tourism Organization (WTO) has initiated a new console on COVID-19 on the tourism sector based on tourism destinations worldwide.

The dashboard incorporates: tourists' entry and receipts, destination vulnerability, assessment of COVID-crisis impact on tourism, and so on.

12.3 FINDINGS AND CONCLUSION

Tourism critically concerns not only what a destination can offer; rather, it reflects how the destination can serve tourist better [18]. In the context of tourist and resident profile, there exist a lot of similarities. The tourist perceives basic amenities like food, transportation, and accommodation as the essential facets while traveling [19]. Importantly, Meghalaya as a tourism destination provokes positive and negative influences [20]. According to Goeldner and McIntosh, "Tourism demand for a particular destination is a function of the propensity to travel and the reciprocal of the resistance of link between origin and the destination areas." Conclusively, a tourist's demand can be defined as "either the number of tourists who will or may visit a particular attraction or region, with a set of expectations and requirements, OR the volume of services, products, and facilities that will or might be purchased from a specified geographic area, for a certain price range, and within a given period of time" [2].

Through this article, the emotions and perceptions of locals and tourists in a crisis were acknowledged. This research focuses on tourist and resident emotions/perception on Meghalaya as a tourism destination. It accounts of both negative and positive feedback, where in some cases the tourists are unable to communicate their problem to the residents and on the other hand the residents are hard to come across to solve the questions owning to language barrier as one of the major differences among them [1]. The tourism sector, which is contemplated as a noteworthy component of many economies, is extremely affected by the pandemic which is affecting the wider economy. The national isolation, quarantine, and lockdowns have enfeebled the demand for tourism and travel abruptly, causing huge income losses and a drastic liquidity stupor for many businesses. There are also many job losses in the tourism sector because of the pandemic. Even when tourism restarts operations, work will run at limited volumes due to new health patterns [21]. Slowly the demand for tourism will recuperate but the longer the pandemic carries on, the more intense the consequences on consumer behavior and travel etiquette will be, leading to major implications [22]. However, tourism is also perceived as a boon in places where job creation is stimulated, and the tourists can also visit places freely regardless of the maturity levels on tourism. These effects could influence residents, and this intimation is also becoming a significant area of the tourism organization process [23]. The perception of the residents and tourists can be pinpointed as being relevant for tourism development or not. This means that a person who receives tourism-related income that is inclined to improve their standard of living is more apparently to have an affirmative perception of tourism development, and will much likely support this sector every time afterwards [24]. However, this might not happen to other individuals, who perceive the drawbacks of the tourism industry, and thus will defy the sector [25]. It was also established that men were more cooperative and encouraged tourism development in the state than women [1]. The main components—namely, positive perception, negative perception, and tourism development—were found out to have an important impact on support for tourism during a crisis. We can conclude that the residents/tourists in Meghalaya trust that tourism development has the potential of yielding constructive as well as fatal effects on the national development in terms of the three facets: socio-cultural atmospheres, the particular environment, and the status of the economy; even in a crisis. Nonetheless, there exists a meaningful state of homogeneity among tourists and residents and their level of perception.

KEYWORDS

- **correlation matrix**
- **cultural heterogeneity**
- **general adaptation syndrome**
- **less developed countries**
- **tourist perception**
- **World Tourism Organization**

REFERENCES

1. Nongsiej, P., & Shimray, S. R., (2017). The role of entrepreneurship in tourism industry: An overview. In: *The National Seminar on Entrepreneurial Opportunities for Educated Youth in Global Business.* Pondicherry University, Puducherry. Retrieved from: https://www.researchgate.net/publication/316240958_The_ROLE_of_ENTREPRENEURSHIP_in_TOURISM_INDUSTRY_An_Overview (accessed on 5 July 2022).
2. Liu, J. L., Sheldon, P. J., & Var, T., (1987). Resident perception of the environmental impacts of tourism. *Annals of Tourism Research, 14,* 17–37.
3. Young, A. A., (1928). Increasing returns and economic progress. *Economic Journal, 38*(152), 527–542.
4. Porter, M. E., (2008). *Competitive Strategy: Techniques for Analyzing Industries and Competitors.* New York: Simon and Schuster.
5. Schumpeter, J. A., (1947). The creative response in economic history. *Journal of Economic History, 7*(2), 149–159.
6. Albu, C., (2013). Stereotypical factors. *Cross-Cultural Management Journal, 15*(1), 5–13.
7. Lankford, S. V., & Howard, D. R., (1994). Developing a tourism impact attitude scale. *Annals of Tourism Research, 21,* 121–139.
8. Cooke, R. M., (1982), Risk assessment and rational decision theory. *Dialectica, 36,* 329–351. https://doi.org/10.1111/j.1746-8361.1982.tb01547.x.
9. Liu, J., & Var, T., (1986). Resident attitudes towards tourism impacts in Hawaii. *Annals of Tourism Research, 13*(2), 193–214.
10. Huang, Y. H., & Stewart, W. P., (1996). Rural tourism development: shifting basis of community solidarity. *Journal of Travel Research, 34*(4), 26–31. https://doi.org/10.1177/004728759603400404.
11. Toma, S. G., Grigore, A. M., & Marinescu, P., (2014). Economic development and entrepreneurship. *Procedia Economics and Finance, 8,* 436–443. https://doi.org/10.1016/S2212-5671(14)00111-7.
12. Brokaj, R., (2014). Local government's role in the sustainable tourism development of a destination. *European Scientific Journal, 10*(31), 103–117.
13. Kanchana, R., Divy, J., & Beegom, A. A., (2013). Challenges faced by new entrepreneurs. *International Journal of Current Research and Academic Review, 1*(3), 71–78.
14. Selye, H., (1956). The Stress of Life. New York: McGraw-Hill Book Company.
15. Pearson, C. M., & Clair, J. A., (1998). Reframing crisis management. *Academy of Management Review, 23,* 59-76.
16. Matthew, W. S., Timothy, L. S., & Robert, R. U., (1998). *Communication, Organization, and Crisis, Annals of the International Communication Association, 21*(1), 231–276. doi: 10.1080/23808985.1998.11678952.
17. Smith, L., & Millar, D., (2002). *Before Crisis Hits: Building a Strategic Crisis Plan.*
18. Doyle, P., (2002). *Marketing Management and Strategy* (4th edn.). Harlow: Prentice Hall.
19. Koh, K. Y., & Haltten, T. S., (2014). The tourism entrepreneur: The overlooked player in tourism development studies. *International Journal of Hospitality & Tourism Administration, 3*(1), 37–41.

20. Meghalaya Entrepreneurship Promotion Strategy, (2020–2025). PRIME (Promotion and Incubation of Market-Driven Enterprises); Planning Department; Government of Meghalaya.

21. Fernandes, P., (2016). *What is Entrepreneurship?* Retrieved from: World wide web: http://www.businessnewsdaily.com/2642-entrepreneurship.html (accessed on 5 July 2022).

22. Ekanayake, E. M., & Long, A. E., (2012). Tourism development and economic growth in developing countries. *The International Journal of Business and Finance Research, 6*(1), 51–63.

23. Loyacono, L. L., (1991). *Travel and Tourism: A Legislator's Guide.* Washington DC: National Conference of State Legislatures.

24. Chen, C. F., & Chen, F. S., (2010). Experience quality, perceived value, satisfaction and behavioral intentions for heritage tourists. *Tourism Management, 31*, 29–35.

25. Butler, R., (1999). Sustainable tourism: A state-of-the-art review. *Tourism Geographies, 1*(1), 7–25.

26. Long, P. T., Perdue, R. R., & Allen, L., (1990). Rural resident tourism perceptions and attitudes by community level of tourism. *Journal of Travel Research, 28*(3), 39.

27. The impacts of tourism and recreational facility development. *The Tourist Review International Journal of Contemporary Hospitality Management, 31*(7), 417–423.

28. Jones, C., (2007). *Developing the Enterprise Curriculum.* Retrieved from: World wide web: http://journals.sagepub.com/doi/pdf/10.5367/000000007783099782 (accessed on 5 July 2022).

29. Mitchell, A., (2015). *Queensland: Knocking off the Hillbilly Dictator: Joh's Corruption Finally Comes Out.* Crikey.

30. National Governor's Association, (1988). *A Brighter Future for Rural America? Strategies for Communities and States.* Washington DC: National Governor's Association.

31. Nemethy, A., (1990). Resorts go up and down. *Snow Country, 3*(7), 31, 32.

32. Chang, J., (2011). Introduction: Entrepreneurship in tourism and hospitality: The role of SMEs. *Asia Pacific Journal of Tourism Research, 16*(5), 467–469. https://doi.org/10.1080/10941665.2011.597572.

33. Oh, C. O., (2005). The contribution of tourism development to economic growth in the Korean economy. *Tourism Management, 26*(1), 39–44. https://doi.org/10.1016/j.tourman.2003.09.014.

34. Othman, P., & Rosli, M., (2011). The impact of tourism on small business performance: Empirical evidence from Malaysian islands. *International Journal of Business, 2*(1), 11–21. https://doi.org/10.5367/000000003101298312.

35. Piotr, D., & Rekowski, M., (2009). *The Relationship Between Entrepreneurship and Economic Growth: A Review of Recent Research Achievements in Entrepreneurship and Business* (pp. 113–136).

36. Bagherifard, S. M., Jalali, M., Jalali, F., Khalili, P., & Sharifi, S., (2013). Tourism Entrepreneurship: Challenges and opportunities in Mazandaran. *Journal of Basic and Applied Scientific Research, 3*(4), 842–846.

37. Programs, B. O., (n.d.). *What Is Entrepreneurship?* Retrieved from: World wide web: https://www.ait.org.tw/infousa/zhtw/DOCS/enterp.pdf (accessed on 5 July 2022).

38. Responsible Tourism Development Fund, (n.d.). *'What Is Responsible Tourism'?* Retrieved from: https://responsibletourismpartnership.org/what-is-responsible-tourism/ (accessed on 5 July 2022).

39. Greiner, R., (2010). Improving the net benefits from tourism for people living in remote Northern Australia. *Sustainability, 2*(7), 2197–2218. https://doi.org/10.3390/su2072197.

40. Smith, A., (1910). *The Wealth of Nations*. Knopf: New York.

41. Taskov, N., Metodijeski, D., Dzaleva, T., & Filipovski, O., (2011). Entrepreneurship in tourism industry lead to business benefits. In: 2nd *Biennial International Scientific Congress: The Influence of Tourism on Economic Development* (pp. 1–12). Retrieved from: http://citeseerx.ist.psu.edu/viewdoc/download?doi=10.1.1.472.4645&rep=rep1&type=pdf (accessed on 5 July 2022).

42. Cobbinah, P. B., Black, R., & Thwaites, R., (2013). Tourism planning in developing countries: Review of concepts and sustainability issues. *International Journal of Social, Behavioral, Educational, Economic and Management Engineering, 7*(4), 1034–1041.

43. Tomazos, K., (2009). *Volunteer Tourism, an Ambiguous Phenomenon: An Analysis of the Demand and Supply for the Volunteer Tourism Market*. Glasgow: University of Strathclyde Glasgow.

44. World Tourism Organization, (2012). *Towards Inclusive & Sustainable Growth & Development: How Can the Tourism Sector Contribute*. Retrieved from: https://www.unwto.org/archive/global/event/towards-inclusive-sustainable-growth-development-how-can-tourism-sector-contribute (accessed on 5 July 2022).

45. World Travel and Tourism Council, (2016). *The Economic Impact of Travel and Tourism*. Retrieved from: https://wttc.org/Research/Economic-Impact (accessed on 5 July 2022).

46. Bryon, J., (2012). Tour guides as storytellers-from selling to sharing. *Scandinavian Journal of Hospitality and Tourism, 12*(1), 27–43. https://doi.org/10.1080/15022250.2012.656922.

47. Echtner, C. M., & Ritchie, J. B., (1991). The meaning and measurement of destination image. *Journal of Tourism Studies, 2*(2), 2–12.

48. Fly, J. M., (1986). *Tourism and Nature: The Basis for Growth in Northern Lower Michigan*. Unpublished Ph.D. dissertation, University of Michigan, USA. Graves.

49. Global Sustainable Tourism, (2014). Retrieved from: www.gstcouncil.org/about/learn-about-gstc.html (accessed on 5 July 2022).

50. Khanka, S. S., (1999). *Entrepreneurial Development*. New Delhi: S. Chand Publishing.

51. Juanita, C. L., Pauline, J. S., & Turgut, V., (1987). Resident perception of the environmental impacts of tourism. *Annals of Tourism Research, 26*(1), 42–42. doi: 10.1177/004728758702600151.

52. John, Ap., (1992). Residents' perceptions on tourism impacts. *Annals of Tourism Research, 19*(4), 665–690. https://doi.org/10.1016/0160-7383(92)90060-3.

53. Iluliana, Ciochina, Carmen Maria, Iordache, & Alexandrina, Sîrbu, (2016). Entrepreneurship in the tourism and hospitality industry. *Management Strategies Journal, 31*(1), 264–275. Constantin Brancoveanu University.

54. Hill, S., (2011). *The Impact of Entrepreneurship Education: An Exploratory Study of MBA Graduates in Ireland*. Retrieved from: World wide web: https://ulir.ul.ie/handle/10344/1663 (accessed on 5 July 2022).

ANNEXURE I

Code	Variable
TF.Q	Tourist food quality
TF.P	Tourist food price
TF.S	Tourist food service
TT.Q	Tourist transportation quality
TT.P	Tourist transportation price
TT.S	Tourist transportation service
TA.Q	Tourist accommodation quality
TA.P	Tourist accommodation price
TA.S	Tourist accommodation service
RF.Q	Resident food quality
RF.P	Resident food price
RF.S	Resident food service
RT.Q	Resident transportation quality
RT.P	Resident transportation price
RT.S	Resident transportation service
RA.Q	Resident accommodation quality
RA.P	Resident accommodation price
RA.S	Resident accommodation service

CHAPTER 13

Are Organizations Ready to Manage Stress During and After the COVID-19 Pandemic?

RAJNI SINGH

Department of Management, Hierank Business School, Noida, Abdul Kalam Technical University (AKTU), Uttar Pradesh, India, E-mail: rajni.singh2009@rediffmail.com

ABSTRACT

COVID-19 is a global health crisis which has not been experienced by anyone in the modern days. Every person has been affected by the pandemic in one way or the other. This study explores the strategies taken up by the organizations and the Individuals to cope up with the challenges posed by the Pandemic. It also explores the Covid-19 stressors responsible for the psychological distress among the employees. It has been found that there is a considerable impact on the behavior of employees due to COVID-19. Stressors like fear of unknown, improper communication, future uncertainty, lack of clarity and role ambiguity, no clarity about the plans to resume work, social isolation are responsible for creating psychological distress among the employees.

13.1 INTRODUCTION

COVID-19 has changed our lives and has forced organizations and employees to change their behavior. COVID-19 is a global health crisis

---

Emotional Intelligence for Leadership Effectiveness: Management Opportunities and Challenges During Times of Crisis. Mubashir Majid Baba, Chitra Krishnan, & Fatma Nasser Al-Harthy (Eds.)
© 2023 Apple Academic Press, Inc. Co-published with CRC Press (Taylor & Francis)

that has not been experienced by anyone in modern days [1]. Every person has been affected by the pandemic in one way or the other. This study explores the strategies taken up by organizations and individuals to cope up with the challenges posed by the pandemic. It also explores the COVID-19 stressors responsible for the psychological distress among the employees. It has been found that there is a considerable impact on the behavior of employees due to COVID-19. Stressors like fear of the unknown, improper communication, future uncertainty, lack of clarity and role ambiguity, no clarity about the plans to resume work, and social isolation are responsible for creating psychological distress among the employees. Psychological distress is responsible for a considerable change in the behavior of the employees and results in prolonged illness. Psychological distress also affects the productivity and performance of the employees. To overcome psychological distress of the employees some of the organizational strategies explored in the study are: (i) more transparency in Communication, (ii) reinforcement of HR policies, (iii) training to mitigate the effects of COVID-19, (iv) social support and workplace relationships, and (v) preparation and communication of proper plan to resume work physically. Some Stress coping strategies must also be developed by the employees to face the challenges of the pandemic. Individuals must devote more time to secure their mental and physical well-being by developing their own stress coping strategies like (i) positive thinking, (ii) prioritizing tasks, (iii) physical fitness, and (iv) spiritual development. Yoga and meditation is a very strong mechanism to cope up with Stressful work-life, it rejuvenates an individual's psychology and clears all the mental blocks. The current study will prepare both Organizations and Individuals ready to handle the challenges thrown by the pandemic.

13.2 METHODOLOGY

The current study is conceptual in nature and is based on literature review. Literature Review helps to examine the current literature to get a comprehensive and complete knowledge about different subjects [2]. Papers of good quality have been reviewed on COVID-19 and psychological stress at the workplace to get a good idea about the measures adopted by the organizations and by the individuals to cope up with the stress due to the outbreak of COVID-19.

13.3 LITERATURE REVIEW

During the pandemic, people faced a number of stressors in their day-to-day life. Government restrictions to move out from home, salary deductions, Job loss, closing organizations and offices, travel restrictions, pressure to keep oneself healthy, accumulating household items to overcome the shortages in the near future were some of the real-life problems which everyone of us might have gone through during this global crisis. COVID-19 is a global health crisis which has thrown economic and social challenges. As a result of COVID-19 a large number of people are under stress. It has been found that prolonged exposure to stress results in severe physical and mental health hazards. Physical problems (fatigue, chronic diseases, reduced stamina to work) and mental problems (depression, anxiety, and stress) may occur due to stress [3, 4]. Reduced productivity and high rate of absenteeism are also the outcome of stress [5].

13.3.1 *ORGANIZATIONAL STRESS COPING STRATEGIES*

Managing stress during and after the pandemic must be taken on high priority by the organizations. In such a scenario, it becomes the responsibility of organizations to take some measures to prevent the potential negative effects of stress on the mind of the employees [6]. In this chapter, some suggestions and recommendations have been given to the organizations so that some mitigation measures can be taken to overcome the outcomes of the pandemic. Human resource (HR) managers must take the lead to adopt some measures to overcome the psychological distress employees are facing due to COVID-19 crisis. Some of the organizational strategies suggested to mitigate the effects of COVID-19 are discussed in subsections.

13.3.1.1 MORE TRANSPARENCY IN COMMUNICATION

Organizational communication plays an important role in overcoming the negative outcomes of COVID-19. Employee's psychological state of mind has become so negative that some clarity in organizational communication is needed to reduce the psychological distress. Managers must have continuous communication with the employees whether they are physically present in the organization or not. Continuous communication prevents social isolation, fear, and uncertainties of their employees. Organizational leadership must be open and honest so that trustworthy relationship can be developed with others [7]. During the COVID-19 crisis, leaders must announce the major policies to all the employees by phone or through video conference, this would help in constructing a shared vision and shared responsibility to work. Transparent communication by the leaders in this time of crisis can reduce some of the stress of their subordinates. Subordinates develop confidence and trust in leadership which thereby impacts subordinate's performance positively [8].

13.3.1.2 REINFORCEMENT OF HR POLICIES

Managers along with HR experts and health practitioners must come up with some health and safety plans to prevent the spread of the COVID-19 disease. HR policies like flexibility plans and teleworking can be reinforced within the organization. Implementation of such policies may make the employees feel more protected and supported by the organization [6]. To control the spread of COVID-19 most of the organizations have asked their employees to work from home by making the use of the internet, email, and phone. Teleworking is one of the best options with the help of which organizations may continue their operations and can also ensure the health, safety, and well-being of their employees both during and after the pandemic. Inadequate psychological support by the employers is responsible for the psychological distress of the employees. HR practitioners must develop some psychological support system for the employees with the help of well-trained mental health professionals who can mitigate the consequences of COVID-19 pandemic [10, 11].

13.3.1.3 TRAININGS TO MITIGATE THE EFFECTS OF COVID-19

Training is a strong weapon to handle the challenges posed by the pandemic. There is a need to train people to handle the challenges of COVID-19. To reduce the psychological distress "stress management training" must be given to all the employees. Initially managers are needed to be trained so that they can understand how to manage the current and aftermath of pandemic. Trained managers can educate and coach their subordinates

Are Organizations Ready to Manage Stress During   195

well about how to handle the challenges of COVID-19. There is a strong need to train Managers about virtual team management and teleworking. Managers will further act as a mentor of their subordinates and will facilitate the desired changes. To reduce the stress level of teleworkers, training must be arranged by the managers to educate them about the effective utilization of the technology. There must also be trainings on COVID-19 to educate people about the desired behaviors and necessary precautions to be taken like sanitation, right eating habits and social distancing to be taken to prevent the spread of the virus. Training is indeed essential both during and after the COVID-19 crisis as it helps in reducing the stigmatization at the workplace [12].

13.3.1.4 SOCIAL SUPPORT AND WORKPLACE RELATIONSHIPS

Social support plays an important role in handling organizational psychological distress [13]. Employers must provide a supportive working environment to the employees in order to mitigate the negative outcomes of quarantine, social isolation, fear of the unknown and uncertainty. Social support system must be developed by the management by ensuring a good communication with the employees with the help of virtual team meetings [14]. Social support can be extended by the organizations by giving employees some flexibility in their work schedule. Gradual return to work may be considered for those employees who were quarantined or has gone through some mental trauma during the pandemic. Social support strengthens the relationship of management with the employees which ultimately reduces their stress level.

13.3.1.5 PREPARATION AND COMMUNICATION OF PROPER PLAN TO RESUME WORK PHYSICALLY

Return to work plan must be well prepared for the employees. It includes the expectations and future plans of the employers. Plans must be prepared for those who were quarantined and were working in an online mode or in teleworking mode. People who resume work after a period of time feel less stressed if there is proper communication about the moves of the organization and about the expectation of the organization from its employees. Organizations can have work accommodation plans for those employees who will resume the work after a long period of quarantine so that mental illness can be subsided easily [15]. Communication of proper plan to resume work physically will reduce the fear of unknown. Such a proper preparation and communication of plan helps in reducing the stigma and mental unrest of the employees [16].

13.3.2 INDIVIDUAL STRESS COPING STRATEGIES

Researches prove that an individual's belief about stress plays an important role in coping with stress in an effective manner [17, 18]. A number of researchers have given different models of stress like transactional model of stress coping [19], biopsychosocial model of stress [20] and recently stress optimization model [21]. All these models of stress suggest that an individual's mindset determines that whether he/she will be able to cope with the stressors in an adaptive manner leading to effective coping or he will follow maladaptive behavior leading to ineffective coping. Maladaptive behavior results into lack of clarity of thoughts and health problems. Hans Selye – the father of Stress management explained that "Stress is inevitable and it is the spice of life" which is needed for optimum performances. Stress is not always unpleasant (distress) but it is also pleasant (eustress). Stress is how our body responds to any threat, change or pressure for some external environment. Stress is created in our mind. Eustress or Adaptive stress is a productive stress which results in better health, correct cognition, high performance, and productivity [18]. Distress or maladaptive stress results in health hazards and improper functioning [23, 24]. For mature coping with stress individuals must have Eustress. Individuals must follow some stress management strategies which can help them to minimize the felt stress and anxiety [25, 26]. During and after the COVID-19 pandemic some stress management strategies are needed by the individuals to overcome the stressors posed by the lockdown, social isolation, job insecurity and other challenges [27]. Some of the individual strategies suggested to mitigate the effects of COVID-19 are discussed in subsections.

13.3.2.1 POSITIVE THINKING

Our mind is responsible for most of our actions. Feed your mind with positive thoughts to face the challenges of COVID-19. Various researches

Are Organizations Ready to Manage Stress During   197

have proved that positive thinking is directly proportional with life satisfaction (LS), well-being, optimism, and overwhelming feelings [28]. Positive thinking is needed to be developed by the individuals to overcome the stress due to pandemic.

13.3.2.2   PRIORITIZING TASKS/PROBLEM FOCUSED APPROACH

Due to COVID-19 most of the people are doing work from home, in such a situation individual finds it very difficult to meet all the personal and professional demands. He/she is required to set priority for all the tasks to be done. Proper time management is the key to relieve most of the stress. First priority must be given to the most important and urgent work later on other tasks can be given time accordingly. If an individual follows an adaptive and problem focused approach then he can resolve most of his stress and will have a higher psychological satisfaction [29, 30].

13.3.2.3   PHYSICAL FITNESS

It is well said that a "healthy mind lives in a healthy body." All individuals must have some fitness mantra for themselves like yoga, exercises, attending gymnasium, running, dancing, etc. During the pandemic, people spend most of their time on laptops, and it is suggested that they must spend at least half an hour to one hour on their fitness. Breathing techniques like Yoga and Pranayam are very helpful in minimizing stress [31].

13.3.2.4   SPIRITUAL DEVELOPMENT

During this time of COVID-19 crisis, it becomes the responsibility of all the individuals to grow themselves spiritually, and this would help in increasing their emotional well-being [32]. Spiritual intelligence helps individuals to develop an understanding of the purpose of life and develops wisdom to make a right choice of action [22]. Spiritual intelligence can be developed through spiritual connect with the supreme power, which can be established through meditation. Individuals must practice meditation daily to avoid the irrational thoughts. Breathing techniques like Yoga and

Pranayam relaxes the mind and help in meditation. Meditation is one of the important keys to solving most of our problems.

13.4 CONCLUSION

This study reveals that the current challenges of COVID-19 are needed to be addressed both by organizations and individuals. The positive psychology of individuals is needed to mitigate the effects of COVID-19. Stressors due to pandemic have created a lot of psychological distress, which ultimately affects the performance, productivity, and quality of work of employees. Employees are needed to have mental well-being, which can be secured by the efforts of both organizations and the employees. Organizational support in terms of supporting HR policies plays an important role in overcoming the stress of the employees [9]. This study will help the managers and employees in developing stress coping strategies both during and after the pandemic.

KEYWORDS

- **correlation matrix**
- **COVID-19**
- **cultural heterogeneity**
- **pandemic**
- **psychological distress**
- **stress coping strategies**
- **stressors**

REFERENCES

1. Kickbusch, I., Leung, G. M., Bhutta, Z. A., Matsoso, M. P., Ihekweazu, C., & Abbasi, K., (2020). Covid-19: How a virus is turning the world upside down. *BMJ, 369,* 1336.
2. Grant, M. J., & Booth, A., (2009). A typology of reviews: An analysis of 14 review types and associated methodologies. *Health Info. Libr. J., 26,* 91–108.

3. Cohen, S., Janicki, D. D., & Miller, G. E., (2017). Psychological stress and disease. *JAMA, 298,* 1685–1687.
4. Kuo, W. C., Bratzke, L. C., Oakley, L. D., Kuo, F., Wang, H., & Brown, R. L., (2019). The association between psychological stress and metabolic syndrome: A systematic review and meta-analysis. *Obesity Reviews, 20,* 1651–1664.
5. Kirsten, W., (2019). Making the link between health and productivity at the workplace: A global perspective. *Industrial Health, 48,* 251–255.
6. Brooks, S. K., Dunn, R., & Amlôt, R., (2018). A Systematic, thematic review of social and occupational factors associated with psychological outcomes in healthcare employees during an infectious disease outbreak. *J. Occup. Environ Med., 60,* 248–257.
7. Hunt, E. K., (2017). Humane orientation as a moral construct in ethical leadership theories: A comparative analysis of transformational, servant, and authentic leadership in the United States, Mexico, and China. *International J. on Lead, 5,* 1–11.
8. Landesz, T., (2018). Authentic leadership and Machiavellianism in young global leadership. *The Journal of International Business, 2,* 39–51.
9. Wallace, J. C., Edwards, B. D., Arnold, T., Frazier, M. L., & Finch, D. M., (2009). Work stressors, role-based performance, and the moderating influence of organizational support. *J. Appl. Psycho., 94,* 254–262.
10. Qiu, J., Shen, B., & Zhao, M., (2020). A nationwide survey of psychological distress among Chinese people in the COVID-19 epidemic: Implications and policy recommendations. *Gen. Psychiatr.,* 33.
11. Zhang, J., Wu, W., & Zhao, X., (2020). Recommended psychological crisis intervention response to the 2019 novel coronavirus pneumonia outbreak in China: A model of West China hospital. *Precision Clinical Medicine, 3,* 3–8.
12. Brooks, S. K., Webster, R. K., & Smith, L. E., (2020). The psychological impact of quarantine and how to reduce it: Rapid review of the evidence. *Lancet, 395,* 912–920.
13. Karasek, R., & Theorell, T., (1990). *Healthy Work: Stress, Productivity, and the Reconstruction of Working Life* (Vol. 66, pp. 525, 526). New York, Basic Books.
14. Greer, T. W., & Payne, S. C., (2014). Overcoming telework challenges: Outcomes of successful telework strategies. *The Psychologist-Manager Journal, 17,* 87–111.
15. Durand, M. J., (2014). A review of best work-absence management and return to work practices for workers with musculoskeletal or common mental disorders. *Work, 48,* 579–589.
16. Bai, Y., Lin, C. C., Lin, C. Y., et al., (2004). Survey of stress reactions among health care workers involved with the SARS outbreak. *Psychiatr Serv., 55,* 1055–1057.
17. Crum, A. J., Salovey, P., & Achor, S., (2013). Rethinking stress; The role of mindsets in determining the stress response. *J. of Personal and Soc. Psychol., 104,* 716–733.
18. Keech, J. J., Cole, K. L., Hagger, M. S., & Hamilton, K., (2020). The association between stress mindset and physical and psychological well-being: Testing a stress beliefs model in police officers. *Psycol. Health, 35,* 1306–1325.
19. Lazarus, R. S., & Folkman, S., (1984). *Stress, Appraisal, and Coping.* New York, NY: Springer.
20. Blascovich, J., & Mendes, W. B., (2010). Social psychophysiology and embodiment In: Fiske, S. T., Gilbert, D. T., & Lindzey, G., (eds.), *The Handbook of Social Psychology* (5th edn.). New York: Wiley.

21. Crum, A. J., Jamieson, J. P., & Akinola, M., (2020). Optimizing stress: An integrated intervention for regulating stress responses. *Emotion, 20,* 120–125.
22. Zinnbauer, B. J., Pargament, K. I., & Scott, A. B., (1999). The emerging meanings of religiousness and spirituality: Problems and prospects. *Journal of Perso., 67,* 889–919.
23. Brooks, A. W., (2014). Get excited: Reappraising pre-performance anxiety as excitement. *J. of Exp. Psychol., 143,* 1144–1158.
24. Crum, A. J., Akinola, M., Martin, A., Fath, S., The role of stress mindset in shaping cognitive, emotional, and physiological responses to challenging and threatening stress. *Anxiety, Stress & Coping, 30,* 379–395.
25. Clough, B. A., March, S., Chan, R. J., Casey, L. M., Phillips, R., & Ireland, M. J., (2017). Psychosocial interventions for managing occupational stress and burnout among medical doctors: A systematic review. *Systematic Reviews, 6,* 144.
26. WHO, (2020). *Coronavirus Disease (COVID-19) Pandemic.* Retrieved from: https://www.who.int/emergencies/diseases/novel-coronavirus-2019 (accessed on 5 July 2022).
27. Knittle, K., Heino, M. T. J., Marques, M. M., Stenius, M., Beattie, M., Ehbrecht, F., & Hankonen, (2020). The compendium of self-enactable techniques to change and self-manage motivation and behavior v1.0. *Nature Human Behavior, 4,* 215–223.
28. Taylor, S. E., Kemeny, M. E., Reed, G. M., Bower, J. E., & Gruenewald, T. L., (2000). Psychological resources, positive illusions and health. *American Psychologist, 55,* 99–109.
29. Fu, W., Wang, C., Zou, L., Guo, Y., Lu, Z., Yan, S., & Mao, J., (2020). Psychological health, sleep quality, and coping styles to stress facing the COVID-19 in Wuhan, China. *Translational Psychiatry, 10,* 225.
30. Rogowska, A. M., Kuśnierz, C., & Bokszczanin, A., (2020). Examining anxiety, life satisfaction, general health, stress and coping styles during COVID-19 pandemic in polish sample of university students. *Psychology Research and Behavior Management, 13,* 797–811.
31. Nes, L. S., & Segerstrom, S. C., (2006). Dispositional optimism and coping: A meta-analytic review. *Personality and Social Psychology Review, 10,* 235–251.
32. Žižek, S. Š., et al., (2012). Connection between the psychical well-being and spiritual intelligence as factors of a requisitely holistic management. *Annals of the Alexandru Ioan Cuza University- Economics, 59,* 185–199.

CHAPTER 14

Apprehension of Organization Culture Through Emotional Stability

GARIMA SAINI[1] and SHABNAM[2]

[1]*Research Scholar, Department of Humanities and Social Sciences, National Institute of Technology, Kurukshetra, Haryana, India, E-mail: Garimasaini3@gmail.com*

[2]*Assistant Professor, Department of Humanities and Social Sciences, National Institute of Technology, Kurukshetra, Haryana, India*

ABSTRACT

This chapter proposes a roadmap which helps in building a culture of continuity which helps in delivering long-term benefits to the organization by recognizing the efforts of all team members, encouraging the voice of employees, prioritising the company's values, and positive and strong connections between team members, continuous learning and organization investing in staff development. An afresh contribution having a new perspective in the field of behavioural sciences and human resources with special reference to the emotional stability of employees and organizational culture. Organizational culture influences the behaviour of the employee directly. This has an effect on decision-making abilities and the skills used in contingency situations. The driving force in employees and innovative behaviours helps in building commitment among employees and promoting innovation in an organization. Creativity and efforts are always put in by the committed employees. Studying organizational culture has focused that employee's performance is concentrated by innovation,

---

Emotional Intelligence for Leadership Effectiveness: Management Opportunities and Challenges During Times of Crisis. Mubashir Majid Baba, Chitra Krishnan, & Fatma Nasser Al-Harthy (Eds.)
© 2023 Apple Academic Press, Inc. Co-published with CRC Press (Taylor & Francis)

stability, consistent attitude and communication which helps in applying organisational culture helps in better understanding customer needs. Due to organizational culture, the employee feels like the team which improves performance [4]. In an organization adapting the new policies, completing tasks, communicating and collaborating helps in problem-solving which is procedural in nature.

14.1 INTRODUCTION

Organizational culture includes the practices, expectations, actions, and collective values of all team members. It can be put up as the collective traits that make the organization what it is. It is created through consistent and authentic behaviors practiced in the organization and is well needed as a prominent key for developing traits needed for business success. Organizations emphasizing on culture make employees feel more comfortable, valued, and supported showing high commitment towards the company. Organizations which understand and are aware of employee's emotions; by what they believe, think, say, decide, controlling emotions in agile conditions, behavior with others and how they recognize as well as control emotions of others. Companies like Microsoft and Salesforce being technology-based companies; an admired brand with world class performers who prioritize organizational culture. Being an important indicator of organizational culture is job satisfaction that is a paramount for almost two-third of employees to stay in the job showing reduced turnover intentions. Organization culture helps in making the company unique by retaining and exercising its culture. It can be done in various ways such as alignment which means pulling together of company's objectives and employee's motivation in one direction; Appreciation is when all the team members are frequently given recognition for contributing in the organization; building trust in organization is vital as it helps in relying on other's in the organization; resilience when included in organizational culture helps leaders in teaching to watch and respond to change; integrity helps the team-members in taking decisions and forming partnership contributing in transparency and honesty in organizational culture; innovation that means organization is using innovative means and applying creativity in every aspect of business; and psychological safety that supports employees and provides a safe environment so a comfortable culture is experienced in the organization. The organizational outcomes like strategic management, decision-making, person-job fit, occupational choice, team building and training programs which are grasped through innovation. Organization productivity depends upon the effective management which ensures coordination and proper job roles among the workers. Efficacious leadership skills and emotional stability in managers like interpersonal skills, relationship building, innovation, flexibility, emotional intelligence (EI), and decision-making helps in meeting the goals, which simultaneously increases the organization productivity and performance of the employees.

This chapter proposes a roadmap which helps in building culture of continuity which helps in delivering long term benefits to the organization by recognizing efforts of all team members, encouraging voice of employee's, prioritize company's values, positive, and strong connections between team members, continuous learning and organization investing in staff development. The leaders understanding emotions of his fellows in respective organization foster an environment that is safe, comfortable for risks, suggesting ideas and giving opinions. This type of environment does not only provide a safe environment to work in collaboration but also wove the organizational culture. Leaders being emotionally intelligent, uses emotions to drive the organization ahead. They have this responsibility to make necessary changes for the organization, and being aware of the fellow employee's emotions regarding the changes, they can prepare and plan the best optimal ways to make them. Organizational culture being the complicated challenge must be owned by government and private agencies which helps in teaching employees the norms, values, references, and guidelines which are implied in the organization. It helps in reducing conflict, unifying employees and motivation, which helps employees in carrying out their duties having a positive impact on their performance and behavior. Strong culture helps in maintaining the performance of the organization in the long run perceiving that all employees have the same mindset which helps in achieving organizational goals. It is important to change the old organizational culture with the new one as it helps in achieving the voluntary desires and participation of employees. Change in the organizational culture will not take place through governance and only change if employees want that consciously and voluntarily. Turnover intention by employees due to change are very few in number. Organizational culture influences the actions and the behavior of the employees [1], being a psychological framework, which is embedded and practiced by the employees [2]. It can be better understood as the personality which influences the manner the employees work within the organization. Research in this field have shown a positive impact of organizational culture on employee's performance [3, 4]. Performance culture will influence the service and quality provided by the organization. Organizational culture influences the driving force in employee's and innovative behaviors, which helps in building commitment among employee's promoting the innovation in organization. Creativity and efforts are always put in by the committed employees. Studying organizational culture has focused that employee's performance is concentrated by innovation, stability, consistent attitude, and communication which helps in applying organization culture helps in better understanding customer needs. Due to organizational culture the employee feels like the team which improves the performance [4]. In an organization adopting the new policies, completing tasks, communicating, and collaborating helps in problem solving which is procedural in nature.

14.2 ORGANIZATIONAL CULTURE

Organizational culture is explained by four levels; (1) Artifacts which add all the occurrence which can be heard, seen, felt as employees discover other cultures and their groups. It has products that adds up language, style, technology, appearance, architecture, and organizational story; (2) embracing beliefs and values – employees are persuaded by the managers to act on the work solutions, and beliefs; (3) having the similar perception of success and values the promotion resulting in trust value that originated through a shared assumption; (4) assumption that nonconfrontational belief of the employees are difficult to change. Though the understandings and beliefs which are difficult to change any approach can be created negatively and interpreted differently [5]. Organizational culture can easily understand through, organizational commitment, emotional stability, job satisfaction. Organizational commitment can be better understood through three dimensions [6]. First, affective commitment among workers will continue doing work as a responsibility with a general feeling of doing more for the organization. Second, a sustained commitment towards the organization. Third, employees having high normative commitment is necessary which leads workers in defending organization despite of facing social pressure. With this it is fair to say that organizational commitment and job satisfaction significantly relates to performance boosting the organizational culture.

Employee have their different behaviors and policies depending on their organizational commitment. Employees wish to become members desire and work in the direction leading their business to meet their organizational goals. Opposite to these are the ones who forcefully become members, avoiding financial losses, trying to do little from their part. Developing the normative component due to socialized experience, depending on how far the employee feels. It helps in creating the sense of obligation in respond to what employees receive from the organization. Job satisfaction is understood by knowing the differences between the needs and expectations the employees possess and what they achieve through work. No difference between their possession and their perception results in satisfaction of the employees. Employee performance is dependent on the organizational culture, organizational commitment, and job satisfaction. Organizational culture in the organization helps in achieving high organizational commitment and job satisfaction [7]. Organizational management is important for the progress and drive growth. It can be achieved through organizational commitment and management commitment. Commitment can be understood by the determination to take decisions which helps in achieving the goals [8]. It also reflects the strong desire of the workers to remain associated with the organization [9]. Therefore, organizations need to give full attention by making employees trust in the organization which will boost their commitment towards the organization [10]. It guarantees corporate survival for the members. Employees who show high commitment sees themselves as the part of the organization in the long run by binding and directing the whole organization [11]. On the other hand, employees who have low commitment sees themselves as outsiders and do not treat them as the member of that organization.

The two components of the commitment are attitude and willingness to act. Attitude is understanding and identification of an organization's goals and the basis of employee's commitment towards the organization. Identification of employees emerges as an attitude which accepts organizational values, organizational wisdom, personal values, and proud member of the organization [12]. The drivers of organizational commitment are the employee satisfaction. High commitment towards the organization are the factors which directly or indirectly drives the performance of the employees. High organizational commitment in employees leads them to be more productive and stable which would benefit the organization [13]. The assessment of organizational commitment and focusing on the higher categories' states that a genuine effort and willingness by employees to carry out the organizational goals [14]. Organizational culture and human resource (HR) management are related to one another and are inseparable in the organization. The HR management manages and impacted the management of the corporate culture. Organizational culture is shaped by the HRs system by influencing workers, salary, incentives, composing executives and implementing strategies.

14.3 HUMAN RESOURCES (HRS): DETERMINATES OF AN ORGANIZATION'S SUCCESS

The determinants of organizational success are dependable on the HRs. HRs and its quality are the driving force which boosts the organization's sustainability, stability, and operations. HRs are managed by organizations for their development and survival [15]. The employees being the organization's corporate assets is important in determining the goals of the organization. In the organization, the goals can be achieved when employees meet the requirements and being task-centered. Every organization strive for improving employee's performance as it is in line of organization achievement. The capabilities of employees are reflected in performance. Optimal performance is marked as good performance. Employee's performance is reflected in organization goals, and corporate leaders should be aware of employee's behavior [16]. Quality of HR is directly proportional to employee's performance. Employee performance correlates with effectiveness, efficiency, and output. Work ethic and high work performance show high correlation with productivity. The high-quality employee requirement is well needed for organizational performance [7].

Organizations need employees to practice functional management practices policies for new practices which are more innovative and adaptive in responding to the rapidly changing environment. Cooperate leaders should note the amount of work performance of employees is cooperated and is the record of results derived from specific employee activities to achieve the goals [18]. Organization is constituted of the individuals and it is a human tendency to experience situations and perceive them as they look. If the employees are dissatisfied, it says that it is a disturbance in acceptance of dissatisfied perceptions. The effective and emotional response which shows various aspects of work, including the set of feelings the employees have about the job. Job satisfaction is the general attitude of the employees towards its job, communication with seniors and fellow workers, follow the rule of the organization by meeting the performance standard. The attitude of the employee towards the job by seeing the difference by how much an employee is earning and how he should earn. When the type and nature of work performed according to the employee's needs, a productive work environment is formed. That may be the reason satisfied employees choose their job situation as compared to the dissatisfied employees. Employees performance in the organization is related his satisfaction at the job. This helps in improving the performance as satisfied employees will perform better and vice versa [17]. Employee performance and job satisfaction are positively correlated as employees tend to perform better as there is a boost in employee performance when there is high job satisfaction. According to preferences and value systems, every individual have different job satisfaction level. The more a particular job meets the need of the employees, the more the level of satisfaction exists [18].

Job satisfaction strongly influences the turnover intentions and the retention [8]. Few factors such as promotion, supervision, and co-workers correlate positively with the job satisfaction. Contextual factors like organizational culture and goal orientation are the personal factors influences the job satisfaction affecting the turnover rate in the organization [19]. Employee's commitment is the attitude which shows the loyalty towards the organization, it is closely related to the psychological aspects such as trust and acceptance in organization goals and values as a result of building trust and loyalty in the organization [20]. Employee relations and organizational commitment when implied together boosts the growth of the organization certainly. The organization's growth and performance are the result of the work of an employee or a team having certain responsibilities and authorities with morals, ethics which helps an organization in reaching the goal [17]. The requirements of the employees which are fulfilled indicates the achieved success at work. This marks work requirement as the potent factor in employee's performance in the organization [21]. Managers, who are aware of their emotions make better decisions, increasing performance and developing employees. HR management with emotional stability helps in inducing positive emotional states and responses. It helps in keeping employees motivated inappropriate solutions which overcome challenges and difficulties on a daily basis.

14.4 EMOTIONAL STABILITY AND ORGANIZATION CULTURE

Emotions are complex with different components like expressive behavior, subjective experience, psychophysiological changes, cognitive process, and instrumental behavior. Emotions can be understood by feelings causing psychological and physical changes influencing behavior. Behavioral tendencies and emotions are correlated. Emotions are known as the driving force behind positive and negative motivation. In an organization, it is well needed that employees are aware of their feelings and are able to monitor them and use them in guiding their own actions and thinking. HRs have the ability to communicate with all level workers in the organization heading it towards effectiveness and success. The emotional stability and its competent skills are potent predictors for organization success, productivity, performance, and leadership [22]. Emotional stability helps in increasing self-awareness, self-knowledge, and knows others. Through Maslow's theory of hierarchy of needs the organizational behavior in the organization is studied with reference to antecedents of organizational culture; job satisfaction and organizational commitment. Malow states that the lowest needs are physiological needs and higher needs are self-actualization. Emotional stability and intelligence in the individual help them in achieving the self-actualization stage [23]. He also added that if lower needs are not satisfied then higher needs never show off. In an organization culture, the environment is maintained when needs like job security, salary, manager, respect toward coworkers and polices are not satisfied, the attention towards oneself and higher needs are not satisfied. When the economic growth is not present in the organization, the low needs are not fundamentally satisfied by the employees. Certainly, dissatisfied lower needs would never turn in meeting the higher level of needs [24]. Emotional stability helps in recognizing and attaining organizational commitment, job satisfaction and reducing stress among the employees. This is beneficial when the lower needs are satisfied. With frequent research, it is found that in an organization and with conversation of employees, when lower needs are not met, the higher needs are not satisfied.

At the time of recruitment, it is very important to give preferences to employees who understand and are aware of emotions and act accordingly. This will help them in recognizing and controlling emotions of others. Emotional and social competencies are necessary for the workplace

Apprehension of Organization Culture   209

helping the employees to adjust in the environment, self-management, interpersonal effectiveness, disagreements, and opinions. Emotions helps in improving the performance playing an important role in productivity and organization's effectiveness. Employee's having high emotional stability are the invaluable assets [25]. Reforming and improvements of traits in the workplace like controlling of emotions and perception which are related to emotional stability are important. Emotional stability being the important predictor helps in ramification of organizational outcomes such as job satisfaction and organizational commitment [26]. A person who understands emotions can make good relationships with bosses, supervisors, and colleagues ultimately results in increasing organizational commitment [27], job satisfaction [28], and job performance [29]. Employees having high emotional stability tend to be more committed to their respective organizations showing high performance, reaches desired goal with job satisfaction [30]. Affective commitment persuades positive emotional stability in employees with job satisfaction and organizational commitment [25]. Keen factors of job satisfaction such as relationship with supervisor, satisfaction with salary, satisfaction with support and policies of organization and satisfaction with organizational culture with reference to overall job directly indicates commitment of the employee [26]. Emotional stability is a powerful tool critical for exceeding goals, improving critical work relationships, and creating a healthy, productive workplace and organizational culture which boosts organizational commitment and job satisfaction. It is important to investigate the effect of emotional stability on management issues and economic problems in an organization. Understanding emotions can help in achieving organizational commitment and job satisfaction.

In a job environment, high emotional stability in employees leads to low occupational stress. Employees having low emotional stability are unaware of the emotions making it difficult to cope up with their feeling ultimately results in having a negative influence on job satisfaction. Employees having high emotional stability put forward that they are aware of their feelings and understand themselves, are adaptive to stress and choose a better coping mechanism [22]. Few researches have put a light on variables such as emotional attention, emotional repair, emotional clarity, and occupational stress, which are associated with emotional stability. High emotional repair and stability boosts the commitment of the employees by lowering the stress levels [14]. Emotional stability can be marked as a potent predictor in organizational outcomes like job satisfaction, organizational commitment, and productivity [26]. Emotional stability leads an employee in understanding his feelings, negative emotions, and controlling stress and frustration, which will help in building a better relationship with supervisors and colleagues increasing organizational commitment and job satisfaction [33]. Research indicates that being emotionally stable in an organization increases the job satisfaction reducing turnover intention [29]. Job satisfaction affect the organizational commitment stating that satisfied employees have higher levels of organizational commitment as compared to employees who are not satisfied. The factors of job satisfaction such as satisfaction with overall job, satisfaction with supervision, satisfaction with salary and satisfaction with policy and support are direct indicators of organizational commitment. Knowing their emotions helps employees in attaining work life balance which makes them feel valued, happier, and less stressful with improved mental well-being, increasing commitment, motivation, loyalties, and decreasing turnover intention.

14.4.1 EMOTIONAL STABILITY: WORK LIFE BALANCE

Globalization and in this era of IT and communication, this century has arisen a various problem for working professionals and one among them in work life imbalance. In in global scenario, the changes such as multi-cultural diversity, working environment and demography changes have made it difficult for an employee to adopt the organization's culture. Emotional stability is employees and work life balance are the need of an hour for current organization culture as they work as a competitive edge in the work environment. Emotional stability in employees is the genetic makeover but like any intelligence, EI can also be developed. Emotional stability improves the employee and organizational performance, playing a significant role in working in the employee and his assistance with colloquies. Employees with emotional stability have capacity, ability, skill, and are self-perceived in identifying, assessing, and controlling emotions which plays a crucial role in modern work style. These emotions also help in evaluating management styles, employee behavior, interpersonal skills, attitudes, and potent potential. It helps employees in organizational chores like planning, job profiling, selection, and recruitment. Emotional stability helps employees in understanding and managing emotions which helps in one's own conduct and relationship with others. Researches have

Apprehension of Organization Culture 211

shown that understanding and managing emotions are potent predictors in gratifying work environment. Work-life balance are the practices in the organization which supports the employees needs and help them in achieving the balance in work life and personal life. Researcher's attention attracted the issue of work life balance for all sectors and all levels of the organization. It defines the satisfactory level of involving the multiple roles in employee life.

In a competitive organization, management of work and family has become challenging. In some in industries like IT, where employees have to work from home and balancing work and family is necessary. Research in this field have suggested organization to ensure a practical and rational work life balance that meets the need of employees and the organization. Work life imbalance in the organizations and not providing the opportunities have increased unproductive and dissatisfied employees, increasing the turnover intentions and attrition. Certainly, building work life balance is not enough but fostering EI and supportive organizational culture is important. The policies that the organization run on are also important. The flexible and creative ideas by the organization are welcomed which increases their well-being and ultimately maximizes the productivity. Research states that employees with higher emotional stability have a good work-life balance. The reason behind this is high emotional stability help them to cope up with the consequences arising out of stress. Employees having low emotional stability are not in a position to handle these stressors well. While working in the groups, high emotional stability will influence the emotions of other employees in such a way that they would maintain a satisfied professional and personal life. The following points helps in better understanding of emotional stability and work life balance in the organization:

- Employees having high emotional stability are more productive, with less stress and low turnover intentions with a balanced persona and professional life.
- Emotions leads employees in regulating of self and other's emotions which will facilitate and motivates the employee increasing their productivity.
- Influencing emotional stability on work life balance dimensions are significantly related, while motivating and expressing emotions which will boost the performance of employee.
- High emotional stability with a satisfactory salary and perks helps in having a better work life balance.

- work-life imbalance is caused in employees if they are not satisfied with the salary resulting in stress and frustration towards family and work.
- A positive high correlation is studied between performance and salary in the form of incentives, hikes, and appraisals, employees who are satisfied with the salary have a good work life balance.
- Emotional stability helps employees in believing for advancement whereas low emotional stability often finds their work as pressure, monotonous, low job satisfaction resulting in work-life imbalance.
- Flexibility of working hours is important as flexible timing helps employees in managing the family ad work responsibility and minimizing the conflict between work and family, improving the performance at work.
- Researches states that flexible arrangements at work would help in attaining a better blending at work; helping the organization in retaining, recruiting, and motivating employees [39].
- Flexible timing at work shows a positive relationship between emotional stability and help in reducing absenteeism, turnover intention, and late comings.
- Understanding the emotional stability of employees which helps in maintaining work-life balance and improving the productivity.

At the lower level of management, emotional stability influences work life balance and shows a positive weak relationship. On the other hand, employees at higher and middle management levels, the higher levels of relationship are studied between emotional stability and work life balance. Certainly, higher stability of emotions in an organization has a positive impact on work life balance [27].

14.4.2 MODERATOR MODEL OF EMOTIONAL STABILITY

Emotional stability is shaped and expresses dispositions reacting in an organization context. This model states that moderating effect of emotion stability is unique and competition. Emotional stability facilitates the self-serving goals which are necessary in an organization and focus employee's advancement [35]. Research also highlights the impact of stressors like dissatisfaction and insecurity causes unfavorable actions by the employees regardless of their awareness of emotions. This model accommodates

Apprehension of Organization Culture 213

the dispositional and organizational factors that have harmful effects on employees having more emotional stability as compared to employees with less emotional stability. Regulations of emotions (ROE) facilitate the achievement of self-serving goals and prosocial behavior. These help in the achievement of goals and helps employees to generate particular emotions that conductively achieves goals in the organization. The associations of prosocial behaviors and prosocial orientation are more in employees having high emotional stability. Emotional stability in employees in avoiding incidental emotions which are unrelated to decisions and hampering the decision-making techniques. The effect of incidental anxiety on risk taking is less in employees with a high ability to analyze the causes and their effect relationship on events happening in their organization. This certainly helps employees in identifying the incidental anxiety which is unrelated to decisions and help them in understanding emotions which will reduce the anxiety. Emotional stability also moderates the association between unrelated judgments and mood induction, helping in practicing a goal directed principle in organization. Emotional stability also helps employees in detecting other's authentic emotions, which helps employee in responding to interpersonal encounters. Research states that knowing emotions determines the psychological wellbeing acting as a constellation of capacities and attitudes often related to emotions.

14.5 PSYCHOLOGICAL WELLBEING: PROFICIENT ORGANIZATIONAL CULTURE

Wellbeing is the capability of employees which help them in actively participating in achieving goals, recreating work and maintain intrapersonal and interpersonal relationship with expertizing in positive emotions [34]. Psychological wellbeing underlines the satisfaction of employees as it is emerging as a satisfaction which emerges as a satisfactory feeling with the individual mental and physical health with interpersonal relations. The wellbeing helps the employee in attaining positive relations, self-acceptance, personal growth, and positive attitude. The connection of psychological wellbeing and emotional stability and a positive relationship is studied. The more the individual is aware of his emotions, more the level of self-esteem and less association with negative emotions like depression [36]. Emotional stability helps employees in building a flexible pattern which brings attitude changes and helping him to come out of failure and neglection. There is a structural chain which exist between the emotional stability and life success. In an organizational culture, the relationship among emotions, personality, and variables of well-being holds an important part in the organization. The employees are evaluated on the basis of social well-being, coping, personality, and emotional stability traits. High emotional stability leading in high mental wellbeing in employees and helps them in achieving self-acceptance, positive relations, personal growth, and autonomy. Employees need encouragement from the higher management and organization which enable them to function and achieve goals in a particular timeframe. High emotional stability leading to better mental well-being helps in improving satisfaction, productivity, and commitment. The special workshops help the employees in knowing self-emotions which alleviate their skills making them emotionally intelligent.

14.6 CONCLUSION

This chapter is an afresh contribution having a new perspective in the field of behavioral sciences and HRs with special reference to emotional stability of employees and organizational culture. Organizational culture influences the behavior of the employee directly. This has an effect on decision making abilities and the skills used in contingency situations. Intrinsic and extrinsic personality factors of the employees impact the emotions and behavior of employees. Organizational culture construes the underlying assumptions, values, beliefs, and interactive ways which contributes in making a unique psychological and social environment in the organization. The expectations of the organization with their employees can be expresses in interactions, inner workings and self-image reflected in employee's job satisfaction, performance, and organizational commitment. Building organizational culture of continuity which helps in delivering long term benefits to the organization by recognizing efforts of all team members, encouraging voice of employee's, prioritizing company's values, positive, and strong connections between team members, continuous learning and organization investing in staff development. Organizations need employees to practice functional management practices policies for new practices which are more innovative and adaptive in responding to the rapidly changing environment. Emotional stability is the keen factor responsibly determines the behavior of the employees in the organization. Understanding emotions helps employees in achieving self-acceptance,

Apprehension of Organization Culture 215

positive relations, personal growth, and autonomy. The emotional stability and its competent skills are potent predictors for organization success, productivity, performance, and leadership. The relationship among emotions, personality, and variables of well-being holds an important part in employee's psychological wellbeing and increasing the capital of the organization. Mental wellbeing of the employees is a potent predictor shaping the employee's interpersonal behavior with executive management level and with colloquies. Public and private sector organization is managing a change in an effective way. Applying emotional stability backs the employees and managers in recognizing and understanding emotions and being emotionally intelligent help them in managing oneself and relations with others. In an organization applying EI includes personnel awareness of emotions, development, teamwork, reacting in environment. It is the responsibility of the organization in coaching employees helping them in performing effectively. Enhancing the skills of employees other than technical skills increases productivity at the job. Stability and knowing emotions help employees in directing and controlling feelings in the organization, boosting job satisfaction, commitment, and productivity at work. These competencies help employees in controlling and managing impulses and mood at job which help in developing emotional maturity, emotional competency, emotional sensitivity, and emotional maturity which are needed in the organization. Effective use of emotions (UOEs) helps in better teamwork which helps in understanding colloquies opinions, ideas, and suggestions resulting in better performance and job satisfaction. Executive in management levels needs emotional stability as they are representing the organization to clients, interacting with people outside and in the organization being role model for employees Managers with high emotional stability understand employees needs by providing constructive feedback. The skills like interpersonal and intrapersonal communication, adaptability, stress management are developed in employees with emotional stability. Emotional stability of employees in different sectors marks a flagpole in communication effectiveness which helps in increasing the capital, manufacturing, and service of the organization. The reality has been that many executives' behaviors are difficult to change and ignore the organizational culture, even though organizational culture is very important for improving employee performance. The bending of employee and employer is a need to have a balanced workforce. Management executive and HR understand the importance of communication for building a good balance between professional and personal life. There should be some changes that must be brought in the organizational culture increasing the productivity, satisfaction, and commitment of employees. The false perception of visibility is equal to productivity should be changed. The organization should concentrate on the effectiveness of meeting the goals rather than the length of working hours. Implementing efficient and effective work life balance programs and policies fosters the culture and understanding their employees better helps in reducing the work-life imbalance. Initiatives must be taken by the organizations to improve the emotional stability of the employees as being emotional intelligent improves employee performance boosting individual and organizational performance. This can be achieved through skillful management and communication helping in increasing productivity. Thus, reducing rigidity in organization culture increases the performance and leads to growth in the organization. Organizing workshop which help the employees in knowing self-emotions which alleviate their skills making the organization a happy and satisfied workforce.

KEYWORDS

- **emotional stability**
- **HR policies**
- **job satisfaction**
- **literature review**
- **methodology**
- **organization culture**
- **organizational commitment**
- **psychological wellbeing**
- **work-life balance**

REFERENCES

1. Ahmetoglu, G., Harding, X., Akhtar, R., & Chamorro-Premuzic, T., (2015). Predictors of creative achievement: Assessing the impact of entrepreneurial potential, perfectionism, and employee engagement. *Creativity Research Journal, 27*, 198–205. doi: 10.1080/10400419.2015.1030293.

2. Asang, S., (2012). Membangun sumberdaya manusia berkualitas: Perspektif organisasi publik. Cetakan ke1. *Surabaya: Brilian International, 8*(9), 67–78.
3. Wahyuni, E., & Puji, M., (2015). *Metode Penelitian Social* (Vol. 34, No. 6, pp. 89–95). Departemen Komunikasi dan Pengembangan Masyarakat.
4. Dewi, I., & Sri, K., (2015). The influence of organizational culture and supervision on performance through the commitment of frontliner employees at Pt Bank Riau Kepri. *Journal of Business Management TPAK, 6*(2).
5. Edison, A., & Yohny, K., (2016). *Human Resource Management.* Cetakan ke-1. Bandung: Alfabeta, McGraw-Hill/UK.
6. Albercht, S. L., Bakker, A. B., Gruman, J. A., Macey, W. H., & Saks, A. M., (2015). Employee engagement, human resource management practices and competitive advantage: An integrated approach. *Journal of Organizational Effectiveness: People and Performance, 2,* 7–35. doi: 10.1108/joepp-08-2014-0042.
7. Fauzi, M., Moch, M., & Warso, A., (2016). Pengaruh budaya organisasi dan kepuasan kerja terhadap kinerja karyawan dengan komitmen organisasi sebagai variabel intervening (Studi Pada Karyawan PT. Toys Games Indonesia Semarang). *Journal of Management, 2*(2), 16–21.
8. Al-sada, M., (2017). Influence of organizational culture and leadership style on employee satisfaction, commitment and motivation in the educational sector in Qatar. *Euro-Med Journal of Business, 12*(2), 163–188. https://doi.org/10.1108/EMJB-02-2016-0003.
9. Colquitt, J., Lepine, J. A., & Wesson, M. J., (2009). *Organizational Behavior: Improving Performance and Commitment in the Workplace* (pp. 169–174). McGraw-Hill/Irwin.
10. Lee, S. Y., & Brand, J. L., (2010). Can personal control over the physical environment ease distractions in office workplaces? *Ergonomics, 53*(3), 324–335.
11. Anderson, C. A., Leahy, M. J., Del Valle, R., Sherman, S., & Tansey, T. N., (2014). Methodological application of multiple case study design using modified consensual qualitative research (CQR) analysis to identify best practices and organizational factors in the public rehabilitation program. *Journal of Vocational Rehabilitation, 41,* 87–98. https://content.iospress.com/articles/journal-of-vocational-rehabilitation/jvr709.
12. Andrew, O. C., & Sofian, S., (2012). Individual factors and work outcomes of employee engagement. *Procedia: Social and Behavioral Sciences, 40,* 498–508. doi: 10.1016/j.sbspro.2012.03.222.
13. Anitha, J., (2014). Determinants of employee engagement and their impact on employee performance. *International Journal of Productivity and Performance Management, 63,* 308–323. doi: 10.1108/ijppm-01-2013-0008.
14. Augusto, L. J. M., Lopez-Zafra, E., Berrios, M. M. P., & Aguilar-Luzon, M. C., (2008). The relationship between emotional intelligence, occupational stress and health in nurses: A questionnaire survey. *International Journal of Nursing Studies, 45*(6), 888–901.
15. Avey, J. B., Wernsing, T. S., & Palanski, M. E., (2012). Exploring the process of ethical leadership: The mediating role of employee voice and psychological ownership. *Journal of Business Ethics, 107,* 21–34. doi: 10.1007/s10551-012-1298-2.
16. Rais, R., (2016). Komitmen organisasi, kepuasan kerja dan lingkungan kerja terhadap Kinerja karyawan Di PT. PLN (Persero) wilayah suluttenggo. *Jurnal Berkala Ilmiah Efisiensi, 16*(1), 45–56.

17. Barasa, L., (2018). International review of management and marketing determinants of job satisfaction and it's implication on employee performance of port enterprises in DKI Jakarta. *International Review of Management and Marketing, 8*(5), 43–49.

18. Rifansyah, O., (2016). *The Influence of Transformational Leadership Style and Organizational Culture on Employee Performance at PT. Bank Rakyat Indonesia (Persero), Tbk. Pekanbaru Regional Office* (Vol. 7, No. 9, pp. 73–82). Department of Business Administration, Faculty of Social and Political Sciences, Riau University, Administrative Sciences Study Program, FISIP, Riau University.

19. Joo, B. B., & Park, S., (2009). Career satisfaction, organizational commitment, and turnover intention the effects of goal orientation, organizational learning culture and developmental feedback. *Leadership & Organization Development Journal, 31–36*, 482–500.

20. Hadian, D., (2017). The relationship organizational culture and organizational commitment on public service quality; perspective local government in Bandung, Indonesia. *International Review of Management and Marketing, 7*(1), 230–237.

21. Sutanto, E. M., (2016). The impact of recruitment, employee retention and labor relations to employee performance on batik industry in Solo City, *Indonesia, 17*(2).

22. Bar-On, R., (2006). The bar-on model of emotional-social intelligence (ESI). *Psicothema, 18*, 13–25.

23. Gibson, J., John, M. I., James, H., & Donnelly, Jr., (2012). *Organizations: Behavior, Structure, Processes.* Boston: McGraw-Hill Companies, Inc.

24. Robbins, S. P., & Timothy, J., (2011). Five-factor model of personality and transformational leadership. *Journal of Applied Psychology, 85*(5), 751–765.

25. Carmeli, A., (2003). The relationship between emotional intelligence and work attitudes, behavior and outcomes: An examination among senior managers. *Journal of Managerial Psychology, 18*(8), 788–813.

26. Daus, C. S., & Ashkanasy, N. M., (2015). The case for the ability-based model of emotional intelligence in organizational behavior. *Journal of Organizational Behavior, 26*(4), 453–466.

27. Shylaja, P., & Prasad, J., (2017). Emotional intelligence and work life balance. *IOSR Journal of Business and Management, 19*(5), 18–21. http://www.iosrjournals.org/iosr-jbm/papers/Vol19-issue5/Version-5/D1905051821.pdf (accessed on 5 July 2022).

28. Wong, C. S., & Law, K. S., (2012). The effects of leader and follower emotional intelligence on performance and attitude: An exploratory study. *The Leadership Quarterly, 13*, 243–274.

29. Guleryuz, G., Guney, S., Aydin, E. M., & Asan, O., (2008). The mediating effect of job satisfaction between emotional intelligence and organizational commitment of nurses: A questionnaire survey. *International Journal of Nursing Studies, 45*, 1625–1635.

30. Nikolaou, I., & Tsaousis, I., (2002). Emotional intelligence in the workplace: Exploring its effects on occupational stress and organizational commitment. *The International Journal of Organizational Analysis, 10*(4), 327–342.

31. Kafetsios, K., & Zampetakis, L. A., (2008). Emotional intelligence and job satisfaction: Testing the mediatory role of positive and negative effective at work. *Personality Individual Differences, 44*(3), 712–722.

32. Sunuharjo, B. S., & Ruhana, I., (2016). The effect of job satisfaction and organizational commitment on employee performance (Studi pada PT. Telekomunikasi Indonesia,

Tbk Witel Jatim Selatan, Malang). *Journal of Business Administration (JAB), 34*(1). Administrasibisnis.student journal.ub.ac.id (accessed on 5 July 2022).

33. Robbins, S. P., (2005). In: Parsayan, A., & Aarabi, S. M., (eds.), *Essentials of Organizational Behavior* (11th edn.). Cultural Research Bureau, Tehran.

34. Hatch, S. L., Feinstein, L., Link, B., Wadsworth, M. E. J., & Richards, M., (2007). The continuing benefits of education: Adult education and midlife cognitive ability in the British 1946 birth cohort. *Journal of Gerontology Series B., 62*, S404–S414. https://doi.org/10.1108/01437731011069999.

35. Kilduff, M., Chiaburu, D.S., & Menges, J. I., (2010). Strategic use of emotional intelligence in organizational settings: Exploring the dark side. *Res. Organ. Behav., 30*, 129–152.

36. Mehmood, T., & Gulzar, S., (2014). Relationship between emotional intelligence and psychological well-being among Pakistani adolescents. *Asian Journal of Social Sciences & Humanities, 3*, 178–185. http://www.ajssh.leena-luna.co.jp/AJSSHPDFs/Vol.3(3)/AJSSH2014(3.3-23).pdf (accessed on 5 July 2022).

37. Paramita, E., Lumbanraja, P., & Absah, Y., (2020). The influence of organizational culture and organizational commitment on employee performance and job satisfaction as a moderating variable at PT. Bank Mandiri (Persero), Tbk. *International Journal of Research and Review, 7*(3), 273–286. https://www.ijrrjournal.com/IJRR_Vol.7_Issue.3_March2020/IJRR0037.pdf (accessed on 5 July 2022).

38. Rahardjo, K. A., (2016). *Pengaruh Motivasi, Kepuasan Kerja dan Komitmen Organisasi Terhadap Kinerja Karyawan Pada PT. Sumber Urip Sejati Di Surabaya* (Vol. 14, No. 2). Media Mahardhika.

39. Christensen, L. B. (2000). *Educational Research: Quantitative and Qualitative Approaches*. Boston, MA: Allyn and Bacon.

CHAPTER 15

Emerging Ethical Leadership in Crisis Management

PAVITRA DHAMIJA[1,2]

[1]*LM Thapar School of Management, Thapar Institute of Engineering and Technology, Patiala (Punjab), India*

[2]*CIDB Centre of Excellence, Department of Construction Management and Quantity Surveying, Faculty of Engineering and the Built Environment, University of Johannesburg, Johannesburg, South Africa; E-mail: pavitradhamija@gmail.com*

ABSTRACT

Ethical leadership has emerged to be one of the significant aspects of crisis management. Leaders are the building blocks in every organization. The implementation and sincere application of this concept can prove extremely beneficial for organizations in the time of pandemics. The existence of different theories has presented leadership from diverse perspectives related to individuals' personality and behaviour. Ethical leadership concept is the theoretical outcome of Social Change Model of Leadership Development by Heri in 1996. This approach aims to highlight the importance of ethics and social responsibility from the perspective of individual, group, and societal domain. Accordingly, to foster leadership development, organizations must try to inculcate two main aspects that includes social responsibility and ethics among contemporary leaders.

Emotional Intelligence for Leadership Effectiveness: Management Opportunities and Challenges During Times of Crisis. Mubashir Majid Baba, Chitra Krishnan, & Fatma Nasser Al-Harthy (Eds.)
© 2023 Apple Academic Press, Inc. Co-published with CRC Press (Taylor & Francis)

15.1 INTRODUCTION

Leaders are the building blocks in every organization. They act as one of the indispensable sources and facilitate organizations to achieve competitive advantage, especially during crisis management. The existence of different theories has presented leadership from diverse perspectives related to individuals' personality and behavior. Over the period of time, most of the organizations across industries have witnessed a revolutionary change towards leadership and its styles. Subsequently, it is becoming difficult for the contemporary organizations to combat with the modern outlook towards leadership and gain effectual results. But, instead of exploring and implementing the modern theories that relate itself to spirituality, ethics, and responsibility, the organizations including educational universities and institutions are sticking to age long traditional approaches. The undignified organizational activities have genuinely questioned the role of corporate ethics, socially ethical leadership, and corporate social responsibility in the contemporary organizations [1]. Such activities in the organizations have indeed confirmed the existence of corruption and amorality, in contrast to the scarcity of socially ethical leadership. The argument that ethical behavior of a leader is of utmost importance in every organization is correct as it inculcates the sense of responsibility among leaders [9]. Accordingly, to foster leadership development, organizations must try to inculcate two main ingredients, i.e., social responsibility and ethics among contemporary leaders.

15.2 THEORETICAL FRAMEWORK OF ETHICAL LEADERSHIP

Ethical leadership concept is the theoretical outcome of Social Change Model of Leadership Development by Heri in 1996. This approach aims to highlight the importance of ethics and social responsibility from the perspective of individual, group, and societal domain. The aspect of responsible leadership has been researched by several researchers in diverse contexts with different names like socially responsible leadership or ethical leadership [2, 10]. The attributed reason for the same can be the wide coverage of elements like ethics, social change, responsibility, etc. However, this chapter will use only ethical leadership to maintain the content uniformity (Table 15.1).

Emerging Ethical Leadership in Crisis Management 223

TABLE 15.1 Social Change Model of Leadership Development

Construct	Description
	Individual Values
Consciousness of self	Awareness of the beliefs, values, attitudes, and emotions that motivate one to take action.
Congruence	Thinking, feeling, and behaving with consistency, genuineness, authenticity, and honesty towards others; actions are consistent with most deeply-held beliefs and convictions.
Commitment	The ability to adapt to environments and situations that are constantly evolving, while maintaining the core functions of the group.
	Group Values
Collaboration	To work with others in a common effort, constitutes the cornerstone value of the group leadership effort because it empowers self and others through trust.
Common purpose	To work with shared aims and values; facilitates the group's ability to engage in collective analysis of issues at hand and the task to be undertaken.
Controversy with civility	Recognizes two fundamental realities of any creative group effort; those differences in viewpoint are inevitable, and that such differences must be aired openly, but with civility. Civility implies respect for others, a willingness to hear each other's views, and the exercise of restraint in criticizing the views and actions of others.
	Societal/Community Values
Citizenship	The process whereby an individual and the collaborative group become responsibly connected to the community and the society through the leadership development activity. To be a good citizen is to work for positive change on the behalf of others and the community.

Source: Adapted from Higher Education Research Institute [12]. A social change model of leadership development: Guidebook version III. College Park, MD: National Clearing House for Leadership Programs.

To understand the nature of the chosen concept, it is very important to understand the concept of ethical leadership from three different perspectives: (i) ethical leader basic roles; (ii) hierarchal ethical leadership; and (iii) relational ethical leadership that a leader needs to play in his/her respective organization.

The three aspects are discussed one by one as under:

1. **Ethical Leader Basic Roles:** The four value-based roles include: (a) visionary – a leader is expected to have a vision/foresightedness

to deal with the unforeseen future organizational situations/issues; (b) citizen – this role directs a leader to assume himself/herself as the citizen of the organization and manage all the public as well as the private matters of the employees; (c) servant – as a servant, a leader must effectively deliver his/her services to the best possible satisfactory level in the organization; and (d) steward – being a steward, a leader acts as a firewall for an organization and protects its valuable resources, for example, tangible, and intangible assets [6] (see Figure 15.1).

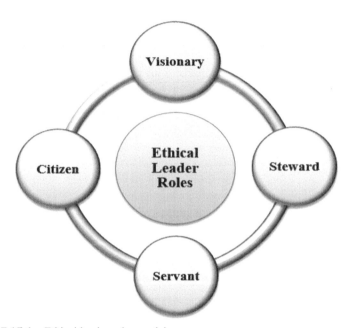

FIGURE 15.1 Ethical leader roles model.
Source: Maak and Pless [7].

2. **Hierarchal Ethical Leadership:** It is a four-level pyramid structure (see Figure 15.2). Originally, the given concept has been coined by Archie Caroll in 1991 as the set of corporate social responsibility features. But, another eminent researcher, Ray Anderson has tried and connected it with the ethical leader concept [4, 5].

To start from bottom to top, (1) an ethical leader must have the quality of being economical. The optimal utilization of funds in the shape of profitability is very much expected; (2) a leader must obey legality in terms of rules, aware of right or wrong codification; (3) an ethical leader is obligatory to do right, just, and fair, and avoids to harm anyone in the organization; and (4) a leader must be a good corporate citizen, contribute resources to community, and endeavor for improve quality of life.

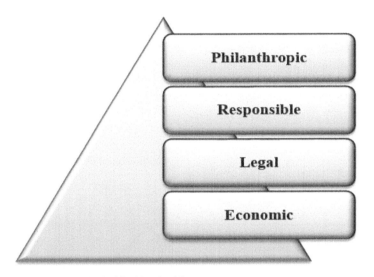

FIGURE 15.2 Hierarchal ethical leadership.
Source: Carroll [3].

3. **Relational Ethical Leadership:** The relational aspect is the extension of hierarchal ethical leadership. It reflects four basic elements, i.e., economic, ecological, socio-political, and moral (see Figure 15.3). Being relations denotes the interdependence of these four elements with each other. A leader is supposed to take into account all these elements while taking any decision in the organization. In a broader view, a leader must consider morality, ethicality, and sustainability before taking any decision [7, 8].

Another aspect that is important to discuss for a far better understanding of the ethical leadership concept is its three vectors [9]. The graphical representation of the same has been shared below (see Figure 15.4).

FIGURE 15.3 Hierarchal ethical leadership.
Source: Carroll [3].

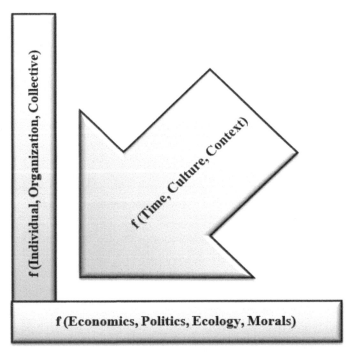

FIGURE 15.4 Ethical leadership vectors.
Source: Miska et al. [8].

Emerging Ethical Leadership in Crisis Management

i. **Ethical Leadership = f (Individual, Organization, Collective):** The first vector denotes that ethical leadership is a function of individuality (me), organizational (we), and collective (us). This denotes the ethical and social responsibility on the shoulders of the leader itself, other staff in the organization and most importantly the stakeholders (consumers, competitors, etc.).

ii. **Ethical Leadership = f (Economics, Politics, Ecology, Morals):** The second vector of an ethical leader is the function of four elements, i.e., economics, politics, ecology, and morals. All these are self-explanatory but are no less in their importance for a leader to deliver efficient and ethical services.

iii. **Ethical Leadership = f (Time, Culture, Contexts):** The last vector confirms that ethical leadership is a function of time, culture, and context. The time involves the amount of time taken to reach a particular level for a leader. It reflects the ethics and morals at specific time (place), culture (nation), and context (social, economic, etc.).

The above section explains the concept of ethical leadership and its importance in different ways. Presently, the concept is being given a lot of weightages by the researchers, due to the increased role of leadership ethics and responsibility in the organizations. An ethical leader can give a competitive edge and sustainability, for which, every organization is struggling presently [10, 11]. The next part shares some more information related to ethical leadership. Ethical leadership denotes that a leader is liable to take responsibility in every situation or crisis. For a leader, a positive and successful task becomes learning, and to accept something which has not met his/her satisfactory standards becomes a responsibility. The aspect of ethical leadership is becoming more and more important among organizations because of its high involvement in achieving organizational effectiveness during pandemic situations. Ethical leadership connotes the control in individuals own behavior by himself/herself. The sense of obligation to be correct with others must be reflected naturally by a leader for the benefit of the organization. The main job of a responsible leader is to develop and maintain cordial relationship with the organizational stakeholders (customers, employees, and business partners) to reach the expected levels of motivation, commitment, and sustainability, which in turn creates value and facilitates social change.

Approaches of ethical leadership differ from each other on moral grounds. The first approach is the normative stakeholder approach. It says

228   *Emotional Intelligence for Leadership Effectiveness*

that the role of ethical leader revolves around the ethicality and altruism. A leader needs to balance the need of stakeholders while keeping in mind the integrity for all. The second approach is related to the economic and strategic aspect, which is extremely important during crisis management. It means that an ethical leader must focus on the wealth maximization and profits in the organization, but with due consideration of values and morals. The selfless intentions of a leader are very important to become an ethical leader. The significance of ethical leadership is very well connected to global pandemic crisis started in 2020. The ethical practices of a leader can be of very much use, if any such crises crop up in the future. The absence of ethicality results in the loss of stakeholders and others concerned with the organization. Pertinently, the purpose of this chapter is to systematically explore, understand, and investigate the major themes related to the concept of ethical leadership in crisis management or pandemic situations.

15.3   EMERGING ETHICAL LEADERSHIP

The available work confirms that ethical leadership has been studied to some extent in different countries of the world. The stated fact can be supported with the scholarly work extracted from Scopus[1] database (1974–2021). The terminology used to access this data includes 'ethical leadership' or 'emerging ethical leadership' as the keywords and all documents published in English are considered. The accessed data (1,387 documents) has been analyzed on various parameters that includes year, subject, country, document, and journal.

15.3.1   YEAR WISE ETHICAL LEADERSHIP

The year wise analysis (see Figure 15.5) confirms that maximum work has been carried out in the year 2020 (197), 2019 (183), 2018 (152), 2017 (130) followed by rest of the years up till 1974. It is evident that over the period of years the concept of ethical leadership has really emerged across the globe. For the present year (2021), 39 documents have already

---

[1] Accessed on 28 February 2021

Emerging Ethical Leadership in Crisis Management 229

been produced, which ensures the importance of this concept during crisis management.

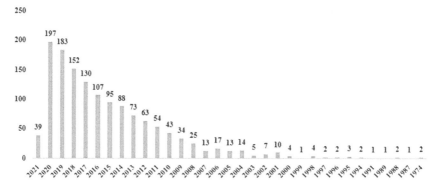

FIGURE 15.5 Year wise ethical leadership.
Source: Scopus data own compilation.

15.3.2 SUBJECT WISE ETHICAL LEADERSHIP

In subject wise analysis (see Figure 15.6), majority of the work has been conducted in the area of business management (444), followed by social sciences (294), economics (145), humanities (104), psychology (100). Other subject areas have also considered this topic for research. The attributable reason can involvement of human as leaders in every organization.

FIGURE 15.6 Subject wise ethical leadership.
Source: Scopus data own compilation.

15.3.3 COUNTRY/TERRITORY WISE ETHICAL LEADERSHIP

the country/territory wise analysis (see Figure 15.7) vividly confirms that the maximum studies are from United States (381), China (114), United Kingdom (90), Australia (79), Canada (70), and followed by other countries. As this chapter explores emerging ethical leadership concept, it is worthwhile to understand that which country is most actively working in this area, especially during crisis management.

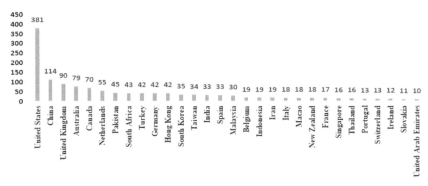

FIGURE 15.7 Country/territory wise ethical leadership.
Source: Scopus data own compilation.

15.3.4 DOCUMENT WISE ETHICAL LEADERSHIP

The document wise analysis (see Figure 15.8) of ethical leadership shows that majority of the work has been conducted in the shape of articles (1096), chapters (101), conference papers (69), review (62), book (21), and followed by other categories.

15.3.5 JOURNAL WISE ETHICAL LEADERSHIP

The journal wise analysis (see Figure 15.9) of emerging ethical leadership confirms the contribution of various authors and researchers. The top listed journal in this category is journal of business ethics (342), leadership quarterly (135), leadership, and organizational development (104), frontiers in psychology (98), business ethics quarterly (98), followed

Emerging Ethical Leadership in Crisis Management 231

by various other journals. These journals can be consulted for advance knowledge of this concept.

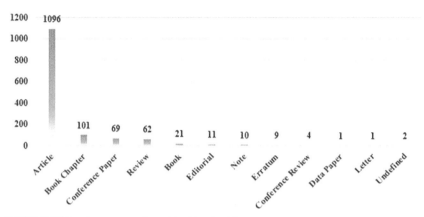

FIGURE 15.8 Document wise ethical leadership.
Source: Scopus data own compilation.

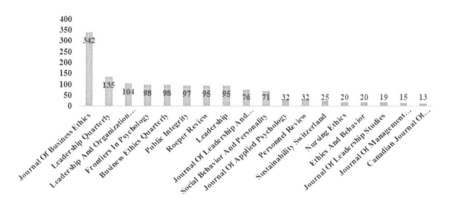

FIGURE 15.9 Journal wise ethical leadership.
Source: Scopus data own compilation.

15.4 RESULTS AND DISCUSSION

The major themes emerged for ethical leadership during crisis management are discussed below (see Table 15.2 and Figure 15.10).

TABLE 15.2 Cluster Classification: Emerging Ethical Leadership

Cluster Number and Label	Current Research	Future Research Suggestions
Cluster 1		
Ethical commitments and organizational performances	Discusses the meaning of ethical commitments and organizational performances. Identifies that ethical leadership is really contribute towards better organizational performances.	Longitudinal studies are recommended to identify further the interconnections between ethical leadership and organizational performances.
Cluster 2		
Ethical leaders and corporate social responsibility	Signifies connection between ethical leaders and corporate social responsibility.	Being ethical is a generalized phenomenon. Combining ethical leadership with corporate social responsibility can bring a fundamental change, which is the need of hour during crisis management.
Cluster 3		
Ethical culture and social justice	Discusses the concept of ethical culture and social justice.	Implementing ethical culture can contribute towards social justice for society at large. Investing resources to bring in ethical culture can is highly recommended.
Cluster 4		
Unethical behavior and workplace deviance	Discusses about the connection between unethical behavior and workplace deviance.	An in-depth analysis of how unethical behavior leads to workplace deviances is recommended.
Cluster 5		
Value-based leadership and workplace spirituality	This theme explains about value-based leadership and workplace spirituality. It evidences workplace spirituality can deliver value-based leadership.	Value-based leadership and workplace spirituality are important aspects to achieve sustainable organizations.

Emerging Ethical Leadership in Crisis Management 233

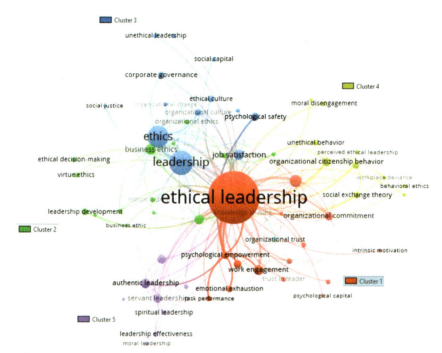

FIGURE 15.10 Cluster analysis for emerging ethical leadership.
Source: Vos viewer visualization.

15.5 CONCLUSION

Ethical leadership has emerged to be one of the significant aspects of crisis management. The implementation and sincere application of this concept can prove extremely beneficial for organizations in the time of pandemics. The various ethical leadership themes emerged from the analysis are highly significant and influential, especially for crisis management when everything has become really unpredictable. The time of crisis sets a very big example that if the leadership is effective everything can fall back to its original and normal state. The concepts like ethics, ethical behavior, trust in leader, ethical decision making, ethical leadership, social justice and other similar aspects form a base for a better and sustainable place to survive in any part of the world. This chapter really contributes towards existing body of knowledge.

KEYWORDS

- consensual qualitative research
- crisis management
- demography
- economic growth
- ethical leaders
- ethical leadership
- globalization

REFERENCES

1. Antunes, A., & Franco, M., (2016). How people in organizations make sense of responsible leadership practices: Multiple case studies. *Leadership & Organization Development Journal, 37,* 126–152.
2. Ardichvili, A., Natt, O. D. K., & Manderscheid, S., (2016). Leadership development: Current and emerging models and practices. *Advances in Developing Human Resources, 18,* 275–285.
3. Carroll, A. B., (1991). The pyramid of corporate social responsibility: Toward the moral management of organizational stakeholders. *Business Horizons, 34,* 39–48.
4. Chebbi, H., Yahiaoui, D., Vrontis, D., & Thrassou, A., (2017). The impact of ambidextrous leadership on the internationalization of emerging-market firms: The case of India. *Thunderbird International Business Review, 59,* 421–436.
5. Christensen, L. J., Mackey, A., & Whetten, D., (2014). Taking responsibility for corporate social responsibility: The role of leaders in creating, implementing, sustaining, or avoiding socially responsible firm behaviors. *The Academy of Management Perspectives, 28,* 164–178.
6. Maak, T., (2007). Responsible leadership, stakeholder engagement, and the emergence of social capital. *Journal of Business Ethics, 74,* 329–343.
7. Maak, T., & Pless, N. M., (2006). Responsible leadership in a stakeholder society–a relational perspective. *Journal of Business Ethics, 66,* 99–115.
8. Miska, C., Stahl, G. K., & Mendenhall, M. E., (2013). Intercultural competencies as antecedents of responsible global leadership. *European Journal of International Management, 7,* 550–569.
9. Muff, K., (2013). Developing globally responsible leaders in business schools: A vision and transformational practice for the journey ahead. *Journal of Management Development, 32,* 487–507.

10. Nonet, G., Kassel, K., & Meijs, L., (2016). Understanding responsible management: Emerging themes and variations from European business school programs. *Journal of Business Ethics, 139*, 717–736.
11. Witt, M. A., & Stahl, G. K., (2016). Foundations of responsible leadership: Asian versus Western executive responsibility orientations toward key stakeholders. *Journal of Business Ethics, 136*, 623–638.
12. Higher Education Research Institute, (1996). *A Social Change Model of Leadership Development: Guidebook Version III*. College Park, MD: National Clearinghouse for Leadership Programs.

Index

A

Absenteeism, 193, 212
Accenture, 28, 142
Adaptive stress, 196
Adventure tourism, 173
Alarm stage, 183
Androgen, 161
Anticipatory quarantines, 83
Anxiety, 4–7, 11, 13, 14, 17, 18, 20–24, 27,
  37, 38, 43, 47, 48, 101, 108–110, 115,
  116, 121–128, 138, 169, 193, 196, 213
Artificial intelligence (AI), 4, 55
Authentic, 23, 107, 109, 111–113, 115,
  116, 223
  behaviors, 202
Automatic conformity, 86
Autonomy, 3, 4, 8, 107, 109, 111, 113, 116,
  214, 215
  depriving event, 109
  restoration, 109
    response, 109

B

Bankruptcy, 82
Behavioral
  sciences, 201, 214
  tendencies, 208
Bio-disasters, 83
Biopsychosocial model (stress), 196
Bivalent logic, 54
Bizarre crisis, 184
Blatantly crisis management, 84
Bodily
  imbalance, 182
  sensitiveness, 175
Branding strategies, 174
Brazil, 37, 41, 43, 47, 48
Brief resilience scale (BRS), 40, 43, 44, 48
British-pay-television company, 144
Burnout, 17, 21, 22, 27, 47

Business
  ethics quarterly, 230
  operations, 19
  progression management, 184

C

Caregiving
  activities, 6
  services, 13
Charisma, 23
Chronic illness, 182
Cisco systems, 144
Civil society, 89
Clinical, 21
  implication (health issues), 53, 54, 56
Cognitive
  dissonance, 22
  dominance, 154
  process, 208
  reappraisal, 25
  skills, 100
Cognizant, 28, 142
Collective lifestyles, 174
Commitment disposition, 123
Common negative emotions (CNE), 163
Communication technology, 13
Community
  infrastructure, 181
  organizations, 174
Compartmentalization, 163
Complex implantable sensors, 131
Computational paradigm, 56
Conflicting factors, 53–57, 70, 77
Conscientiousness, 23
Consensual qualitative research, 234
Conservation of resources (COR), 24
Constellation (capacities), 213
Constructive
  accountability, 92
  feedback, 215

Contagious negative emotions, 25
Containment zones, 108
Contingency situations, 201, 214
Conventional
  channels, 82
  technology, 54
Coronavirus (CoV), 2, 17, 18, 20, 21,
  27–29, 48, 82, 84, 85, 88, 90, 94, 108,
  109, 121–128, 135, 162, 170
  disease, 18, 29
    2019 (COVID-19), 1–9, 12–14,
      17–24, 26–29, 37–41, 43, 45–48,
      81–93, 99, 100, 107–109, 111,
      113–116, 122, 128, 131, 132,
      134–138, 141–145, 147, 155, 161,
      164–168, 170, 175, 176, 178, 184,
      185, 191–198
    Care Centers (CCC), 28
    pandemic, 22, 26, 47, 83, 85, 95, 115,
      116, 131, 170, 194
  infected
    patients, 127
    people, 122, 123, 125–128
  outbreak, 28, 85, 121, 127
  pandemic, 90, 122
Correlation
  coefficients, 180
  matrix, 180, 186, 198
Counseling facilities, 11, 13
Creative, 90, 91, 95, 161, 164, 202
  collaborative strategies, 91
Crippled economies, 28
Crisis, 17–26, 28, 29, 37, 38, 53, 54, 56, 76,
  81–92, 94, 108, 131, 132, 134–138, 161,
  162, 178, 180, 183–186, 191, 193–195,
  197, 221, 222, 227–231, 233, 234
  management, 23, 92, 183, 184, 221,
    222, 228–231, 233, 234
Critical work relationships, 209
Cross
  sectional correlations, 112
  sector collaboration, 85
Cultural
  attractiveness, 177
  heterogeneity, 177, 186, 198
  values, 182

D

Data
  analysis, 111
  collection, 145, 178
Decision-making
  capacity, 161
  character, 160
  techniques, 213
Decorum, 175
Demographic, 210, 234
  information, 110
Depression, 4–7, 11, 13, 14, 17, 18, 20–23,
  27, 37, 38, 43, 47, 101, 108, 110, 116,
  121–128, 160, 162, 163, 166–169, 193,
  213
  anxiety stress scale (DASS), 121, 124,
    126–128
  symptoms, 22, 122
Destination
  attributes, 174
  cleanness, 177
  imagery, 177
    compositions, 177
  impression, 177
  perception, 174
  reasonableness, 177
  vulnerability, 185
Detoxification, 135
Diagnostic systems, 131
Digital social media, 21
Discretionary disposable income, 173
Disruption measurement, 26
Distress, 2, 5, 22, 43, 45, 47, 48, 115, 170,
  191, 192, 196
Domestic travel, 176

E

Ebola, 2, 85
E-conferencing, 144, 155
Economic, 108, 227, 229
  activity, 22
  disruption, 83
  growth, 2, 208, 234
  momentum, 2
  organizations, 178
  performance, 176

Index 239

Education
  establishments, 83
  institutions, 108, 165
  loss, 53, 54, 56, 77
  universities, 222
Effective
  communication, 39
  intervention (health care providers), 136
  prevention policies, 42
  sensemaking skills, 23
Ego orientation, 8
E-health care, 131
Embedded technology extended (ETX), 144
Emergency
  financial assistance, 28
  health services, 93
  packages, 28
  services, 165
Emerging ethical leadership, 228
  country-territory wise ethical leadership, 230
  document wise ethical leadership, 230
  journal wise ethical leadership, 230
  subject wise ethical leadership, 229
  year wise ethical leadership, 228
Emotion, 21–25, 29, 54, 99, 101, 105, 159–163, 169, 170, 175, 212
  attention, 209
  balance, 141, 143, 145, 146, 155
  chaos, 110
  clarity, 209
  competency, 215
  consequences (pandemic), 17
  contagion, 25
  control, 39
  crisis situations, 19, 26, 29
  dissonance, 22
  disturbance, 81
  exhaustion, 21, 22
  fluctuations, 17–19, 23, 26
  health (employees), 23
  imbalance, 146
  intelligence (EI), 1, 17, 23, 25, 26, 28, 29, 32, 37–40, 42–48, 53–57, 69, 70, 76, 77, 81, 99–102, 104, 105, 107, 121, 131, 141, 159–161, 173, 191, 201, 203, 210, 211, 214–216, 221, 234
  inventory, 48

  scale (EIS), 40
  knowledge, 54, 55
  labor, 160
  laden events, 25
  leadership, 20
  management, 23
  maturity, 215
  participation, 154
  quotient (EQ), 53–55, 76, 93, 100, 101, 104
  ramification, 108
  reality, 39
  regulation (ER), 24, 25, 29
  relationships, 154
  repair, 209
  responses, 22, 38, 173
  sensitivity, 215
  stability, 23, 37, 38, 201, 203, 204, 207–216
    organization culture, 208
    traits, 214
  strains, 22
  stress (pandemic), 5
  swings, 18
  turmoil, 24
  work capability, 163
Empirical findings, 105
Employee
  assistance policies, 27
  commitment, 205, 207
  emotional
    balance, 141
    health, 19, 29
  interpersonal behavior, 215
  mental disorders, 28
  motivation, 202
  perception, 141, 142, 155
  performance, 23, 25, 201, 205, 204, 206, 207
  productivity, 92
  psychological
    health, 22, 24, 28
    state, 193
  requirement, 206
  sustainability, 154
  vulnerable emotional states, 19
Enhanced
  healthcare processes, 81

240 *Index*

thinking, 100
Entertainment services, 177
Environment
  degradation, 154
  stresses, 39, 123
  sustainability, 155
Epidemic, 23, 38, 47, 54, 87, 141–143, 145, 155
Epidemiological characteristics, 128
Epinephrine, 161
Essential service vehicles, 165
Esthetic needs, 8
Estrogen, 161
Ethical, 225, 228
  decision making, 233
  leaders, 234
  leadership, 221–223, 227, 228, 232–234
    concept, 221, 222
Exacerbation, 135
Executive management level, 215
Exhaustion stage, 183
Experience sampling method (ESM), 110, 113
Extraversion, 23, 91
Extreme negative contagion, 25
Extrinsic personality factors, 214

F

Facebook, 28, 88, 145
Face-to-face interviews, 145
Family relationships, 174
Fast-changing technology markets, 90
Financial
  disruptions, 26
  instability, 22, 28
  sector, 142
  security measures, 28
  setbacks, 18
  Times Stock Exchange (FTSE), 86
Foreign exchange benefits, 181
Fortune Harvard business review (FHBR), 161
Functional management practices policies, 206, 214
Fuzzy logic, 55, 57
  controllers, 55
  system, 77

G

General adaptation syndrome (GAS), 182, 183, 186
Global
  alliances, 85
  economy, 162
  emergency, 108, 116
  health, 91, 135, 161, 162, 170, 184, 191, 193
  investors, 142, 155
  network, 85
  policy initiatives, 142
  tourism industry, 177
Globalization, 84, 94, 210, 234
Goal-oriented behaviors, 107
Godrej, 28
Google, 20, 28, 39, 109, 132, 142
Government
  emergency response team, 86
  mandated
    closure regulations, 27
    norms, 18, 20
    restrictions, 193
Grand-mean centered, 111
Grocery stores, 21
Gymnasium, 197

H

Hand sanitizers, 82, 90, 142
Health
  care
    education, 133
    facilities, 28
    industry, 82
    planning, 93
    providers, 1–4
    science, 2
    systems, 88
    workers, 4, 5, 12, 137
    workforce, 6, 7
  complications, 162
  conducive working environment, 137
  habits, 135
  infrastructure, 29
  monitoring, 131
  preventing behavior, 123

Index 241

professional sample, 45, 47
programs, 135
related problems, 132, 133
safety organizations, 141
virtual interactions, 136
Hierarchal ethical leadership, 224
Hindustan Computers Limited (HCL), 28
Homogeneity, 178, 186
Hospitals, 6, 12, 13, 21
Hostility, 159
Household chores, 115, 135
Human
  intervention, 55
  need requirement, 8
  resource (HR), 19, 26, 142, 162,
    192–194, 198, 201, 206–208, 214,
    215
    information system (HRIS), 142, 155
    policies, 192, 194, 198, 216
  sensory organs, 54
Hunger of,
  stimuli, 3
  strokes, 3
  structure, 3
Hygiene, 134
  practices, 41
Hypothalamus hypophysial portal axis, 161

I

Immense emotional turmoil, 18
Impaired cognitive functions, 133
Improved clinical processes, 91
Individual stress coping strategies, 196
Indo-Asian News Service (IANS), 144
Indoor
  games, 135
  hobbies, 135
Industrial
  economic stability, 161
  growth, 161, 162
  workplaces, 161
Indwelling premorbid personality, 133
Infectious environment, 47
Infodemic, 21
Information technology (IT), 142, 176,
  210, 211
Infosys, 28
Inhabitants interactions, 182

Innate attractions, 177
Innovation, 81–84, 90–93, 95, 174, 182,
  201–204
  catalyst, 92
  crisis, 95
  strategies, 143
Insomnia, 5, 38, 121, 122, 182
Inspirational motivation, 24
Instagram, 88
Instrumental behavior, 208
Intangible assets, 224
Integrity, 23
Intellectual
  stimulation, 24
  transmittal, 177
Intelligence, 23, 38, 45, 46, 55, 77, 93,
  100, 160, 197, 208, 210
  quotient (IQ), 53, 76, 77, 93
Interconnected dispositions, 123
Internal consistency, 40, 41
International
  organizations, 20
  tourism, 176
Interpersonal
  effects, 53, 54, 56, 77
  encounters, 213
  factors, 133
  relations, 213
  relationship, 39, 100, 162, 213
  skills, 203, 210
Intra class correlation coefficient (ICC),
  111, 112, 116
Intrapersonal communication, 215
Intuitionistic
  fuzzy logic, 53, 55–57, 76, 77
  logic system, 76
Irritability, 182

J

Job
  designation, 110
  enrichment, 159
  insecurity, 22, 26, 196
  performance, 25, 163, 209
  profiling, 210
  satisfaction, 202, 204, 205, 207–210,
    212, 214–216
Judgmental sampling methods, 6

K

Kessler psychological distress scale, 40, 43, 45, 46, 48
Knowledge perception, 55, 56

L

Leadership, 18–20, 23–29, 47, 82–92, 94, 160, 164, 194, 203, 208, 215, 221–223, 225–231, 233
  amid pandemic crisis, 95
  crisis, 95, 228
  development, 221–223
Less developed countries (LDCs), 182, 186
Life satisfaction (LS), 99, 101–105, 197
Linear regression, 99, 104, 105
Liquidity stupor, 185
Local
  associations, 83
  healthcare workforce, 93
  self-isolation, 85
Logical reasoning ability, 55

M

Maintenance price, 177
Maladaptive
  behavior, 196
  stress, 196
Mamdani-type fuzzy logic controller, 57
Management
  commitment, 205
  emotion, 101, 105
  executive, 215
Managerial elasticity, 184
Maslow hierarchy need theory, 154
Mathematical computation, 53
MATLAB, 57, 70, 77
Mediative fuzzy
  linguistic variables, 76
  logic, 53, 55–57, 70, 76, 77
    conflict, 77
  system, 76
Medical
  capacity, 176
  devices, 81
  resources, 5
  technologies, 90, 91

workforce, 1–6, 8–14
  counseling services, 13
  psychological stress, 13
Meditation, 137, 198
  mindfulness, 137
Mental
  disability, 162
  health, 5, 22, 27, 29, 38, 43, 48, 121, 122, 127, 132, 137, 193, 194
  challenges, 122
  inhabitants, 127
  illness, 37, 38, 122, 123, 196
  stability, 163
  support decisions, 5
  toughness, 124
Metaphorical wings, 91
Methodology, 155, 216
Middle East respiratory syndrome (MERS), 2, 14, 82, 95
Mobile phone network, 136
Modern digital technology, 136
Modes of transmission, 134
Mood
  disorders, 109
  induction, 213
Morbidity, 132, 133
Mortality
  rate, 122
  tolls, 20
Motivational inspiration, 162
Multi-cultural diversity, 210
Multi-level
  analyzes, 112
  regression model, 111
Multiple medical comorbidities, 133

N

Narcissism, 86
National
  development, 186
  interest, 88
  isolation, 185
Nationwide complete lockdown, 108, 116
Natural
  catastrophes, 184
  reserves, 177
  scenery, 177

Index 243

Negationism, 48
Negative
emotions, 159, 163
perception, 186
Neurotic personality, 113
Neuroticism, 107, 109, 111–113, 115, 116
characteristics, 109, 116
index, 109
interaction, 111, 113
personality characteristics, 107
New information communication technologies (NICT), 175
Non-cognitive skills, 100
Non-communicable diseases, 132, 133
Non-confrontational belief, 204
Non-essential services, 41
Non-financial disruptions, 26
Non-membership
functions, 57, 69
values, 57
Non-somatic pain, 38, 39
scale, 40, 43, 44
Norepinephrine, 161
Novel coronavirus (nCoV), 2, 17, 18, 29, 82, 84, 87, 108, 131, 176, 177

O

Obsessive-compulsive
disorder (OCD), 23, 29
neurosis, 110
symptoms, 5
Occupational stress, 209
Omnicom Media Group (OMG), 144
Online apps, 145
Optimal performance, 206
Optimism, 24, 138, 197
Organization
changes, 141
commitment, 26, 153, 154, 204, 205, 207–210, 214, 216
communication, 193
crisis, 94
culture, 201–211, 213–216
development, 230
effectiveness, 209, 227
goals, 160, 203, 205, 206

growth, 23
investment, 27
management, 205
moral requirements, 184
performance, 22, 206, 210, 216
productivity, 27, 203
shock waves, 175
stakeholders, 227
strategies, 192, 193
stress coping strategies, 193
structure, 177
sustainability, 206
wisdom, 205
Others emotion assessment (OEA), 99, 101, 103–105
Overt behaviors, 159
Oxford Happiness Questionnaire (OHQ), 40, 43–46, 48

P

Pandemic, 5, 6, 9, 12, 17–23, 26–29, 37–43, 45–48, 53–57, 70, 76, 77, 81–85, 87–94, 99–101, 107–109, 115, 116, 122, 128, 131–133, 135, 137, 138, 147, 161, 164, 166–168, 170, 175, 178, 180, 184–186, 191–198, 221, 227, 228, 233
affected
emotional outcomes, 20
factors, 18
layoffs, 22
organization, 24
provoked employee emotional
experiences, 18
situation (COVID-19), 6
Panic attacks, 37, 38
Paracetamol, 82
Paramedical staff, 1, 3
Participant, 110, 124, 166
consent form, 40
Paytm, 142
Penchant, 175
Perceptual crisis, 184
Performance, 23, 24, 38, 43, 86, 87, 90, 91, 100, 127, 159–161, 163, 164, 192, 194, 196, 198, 202–209, 211, 212, 214–216
oriented team, 91

Index

Personal
competencies, 39, 100
growth, 3, 11, 213–215
Pharma industries, 160, 169
Pharmaceutical
company, 166
fields, 90
industry, 160, 161
manufacturing
factory, 164
industries, 160, 164, 165
workers, 170
sectors, 160
workers, 161
Pharmacies, 21
Physical
activity, 135
distancing, 4, 110
fitness, 192, 197
Physiological
arousal, 159
behaviors, 160
Policymakers, 28, 182
Positive
emotion (PM), 88, 160, 169
climate (PEC), 161
connections, 174
experiences, 173
perception, 186
thinking, 192, 196, 197
Post-crisis growth, 91
Post-traumatic stress disorder, 47, 122
Practical
assistance, 137
training (employees), 10
Pranayam, 197, 198
Precautionary
confinement, 18
measure, 85
Presence Across Nation (PAN), 145
Preventive behavioral measures, 41
Prioritizing tasks, 192
problem focused approach, 197
Private
agencies, 203
sector organization, 215
Probability theory, 54

Productive
organizations, 26
work environment, 207
Professional
health systems, 86
needs, 133
network, 11
Program computation, 76
Prosocial
behavior, 213
orientation, 213
Protective equipment, 1, 3, 5, 10, 12, 13
tools, 13
Psychiatric
hardiness, 122
society, 122
Psychoeducation, 134, 138
Psychological, 4, 22, 192, 198, 229, 230
behavior, 53, 54, 56, 76, 77, 161
changes, 1, 3
deformities, 161
development, 3
disorders, 5, 6, 86
distress, 4, 5, 40, 41, 45, 47, 191–195, 198
factors, 12, 13, 39, 170
growth, 3
hardiness, 121–123, 126–128
scale, 124
health, 20, 24, 26, 29
integrity, 8
needs, 1–6, 8–13
assessment, 5
caregivers, 1, 3, 4
predisposition, 132
problems, 4, 5
process, 175
resources, 28
responses, 5, 7, 9–12
retrieval, 107–113, 115, 116
process, 107–113, 115, 116
safety, 202
satisfaction, 197
stress, 38, 39
related disorders, 1, 4
support medical workforce, 13
symptoms, 6, 47
toughness, 124, 126, 127
score, 124
wellbeing, 109, 213, 215, 216

Index 245

Psychophysiological changes, 208
Psychosocial, 47, 133
   factors, 133
Public
   health systems, 2, 143
   sector entities, 85
   transit, 176

Q

Quality work, 161, 198
Quantitative study, 164
Quarantine, 18, 41, 87, 110, 122, 131, 134, 143, 185, 195, 196

R

Radical innovations, 174
Random-intercept slopes multilevel models, 111
Rationales (destination fealty), 177
Real-time notifications, 21
Recruitment process, 144
Regulation of emotions (ROE), 99, 101, 103–105, 213
Relational ethical leadership, 225
Relationship management, 100
Religious programs, 135
Research
   design, 110, 145
   methodology, 6
      data analysis, 7
      fuzzy logic controller (rule system), 57
      fuzzy logic, 56
      intuitionistic fuzzy logic, 56
      mediative fuzzy logic, 57
      research design, 6
      research instruments, 6
      sample selection, 6
Resident
   host perceptions (tourism), 174
   perception, 178–180
Resistance stage, 183
Resources availability, 178
Respiratory hygiene, 134
Resurrection plans, 178
Rule inference model, 55

S

Salary deduction, 100, 193
Sampling dataset, 107
Satisfaction with Life (LS), 99
   scale (SWLS), 102
Self-emotion appraisal (SEA), 99, 101, 103–105
Self-acceptance, 213
Self-actualization, 3, 208
   stage, 208
Self-awareness, 23, 25, 39, 54–56, 58, 59, 100, 160, 208
Self-care, 132–134
   activities, 133
   intervention plan, 132
   maintenance, 134
   management, 132, 134, 138
   monitoring, 134
   policy programs, 133
   strategy, 134
Self-confidence, 25
Self-efficacy, 23
Self-emotion appraisal, 101, 105
Self-management, 55, 56, 60, 61, 132–134, 138, 209
   program, 134
   strategies, 134, 138
Self-monitoring, 23
Self-motivation, 55, 56, 64, 65
Self-operating program, 57
Self-quarantine measures, 134
Self-serving
   agendas, 88
   goals, 212
Semi-structured interviews, 6, 11
Sensatory ideas, 175
Severe acute respiratory syndrome (SARS), 2, 14, 82, 84, 85, 108, 122
   coronavirus 2 (SARS-CoV-2), 2, 84, 108
Sheer efficacy, 138
Shopping malls, 108, 165
Simulation technique, 76
Skill development, 160
Smoldering crisis, 184
Snowball sampling techniques, 110
Social
   awareness, 100
   competencies, 39, 100, 208

disruption, 54, 182
distancing, 2, 18, 41, 85, 87, 116, 195
environment, 214
ethical leadership, 222
gatherings, 134, 142
interactions, 163
isolation, 20, 37, 38, 41, 110, 191, 192, 194–196
justice, 233
media, 18, 20–22, 40, 88, 135, 136
platforms, 88, 135, 136
psychosocial effects, 47
responsibility, 184, 221, 222, 224, 227
sciences, 229
support workplace relationships, 195
well-being, 137, 214
Socio-cultural atmospheres, 186
Socio-demographic survey, 41
Socioeconomic
classes, 133
crisis, 184
Socio-occupational dysfunction, 133
Socio-political, 225
Software calculations, 76
Solution-oriented technologies, 90
Spiritual
development, 192, 197
intelligence, 197
Staff
development, 201, 203, 214
realization, 184
Stakeholders, 28, 81, 84, 89, 181, 183, 227, 228
Statistical
software, 111
treatment, 125
Stigma, 136
Stigmatization, 195
Strategy, 88, 124, 132–134, 138, 154, 155
Stress, 1, 4–7, 9–11, 13, 14, 17, 18, 20, 21, 23, 24, 26, 28, 37, 38, 41, 43, 47, 48, 54, 99–101, 105, 108, 110, 115, 121–128, 135–137, 143, 159–161, 163, 169, 170, 182, 183, 192–198, 208–212, 215
busters, 135
coping strategies, 192, 198
crisis management, 182
hormones, 182

management, 196
strategies, 196
situations, 159, 160
tolerance, 123
Stressors, 22, 123, 191–193, 196, 198, 211, 212
Sudden crisis, 184
Superspreader events, 88
Supply chain breakthroughs, 91
Sustainability, 225, 227
Symptomatology, 43
Synergetic manner, 85

T

Target deadlines organizations, 147
Task orientation, 8
Technological, 55, 93, 131, 177
companies, 202
Teleworking, 194, 195
Tension, 17, 18, 20, 122, 126, 182
Theorization (emotions), 173
Therapy sessions, 27
Time-location, 145
Tour operators, 176
Tourism, 173, 176, 177, 182, 185, 186
business entity, 176
destination, 177, 180, 185
development, 174, 177, 178, 182, 186
expansion, 182
industry, 173, 177, 186
mechanisms, 176
organization process, 186
planning, 174
programming, 182
related income, 186
sector, 174, 176, 185
Tourist
destinations, 173, 174
gratification, 177
perception, 178, 179, 186
Traditional
attractions, 177
ceremonies, 174
fuzzy logic, 56, 57
intuitionistic fuzzy logics, 56
Traffic jams, 154
Transactional analysis theory, 3

Index

247

Transformational leadership (TL), 24, 25
Transition magnitude, 82
Transparent communication, 194
Travel
  frustrations, 154
  restrictions, 176, 185, 193
Turnover intention, 210, 212

U

Unemployment, 22, 92
Unhealthy diets, 132
Unparalleled intensity, 82
Unprecedented
  economic catastrophe, 17
  universal travel limitations, 176
Upgraded healthcare processes, 90
Use of emotion (UOE), 99, 101, 103–105,
  215

V

Ventilators, 82, 87, 90
Video conference, 194
Viral pandemics, 2
Virtual
  conferencing, 109
  consultations, 136
  team
    management, 195
    meetings, 195
Virus-infected people, 20

W

Washington financial industry, 142
Wealth maximization, 228
WebEx, 109
WhatsApp, 145
Wipro, 28, 142
Work
  accommodation plans, 195
  ethic, 206
  life
    balance, 9, 10, 21, 141, 143, 146, 147,
      155, 210–212, 216
    imbalance, 210, 211
  performance motivation training, 164
  place vehemence, 184
World
  Health Organization (WHO), 2, 18, 21,
    29, 41, 82, 122, 134, 136
  Tourism Organization (WTO), 185, 186

Y

Your objectives guidelines assessment
  (YOGA), 135

Z

Zoom, 109
Zoonotic diseases, 83